T0137202

Studies in Computational Intelligence

Volume 838

Series Editor

Janusz Kacprzyk, Polish Academy of Sciences, Warsaw, Poland

The series "Studies in Computational Intelligence" (SCI) publishes new developments and advances in the various areas of computational intelligence—quickly and with a high quality. The intent is to cover the theory, applications, and design methods of computational intelligence, as embedded in the fields of engineering, computer science, physics and life sciences, as well as the methodologies behind them. The series contains monographs, lecture notes and edited volumes in computational intelligence spanning the areas of neural networks, connectionist systems, genetic algorithms, evolutionary computation, artificial intelligence, cellular automata, self-organizing systems, soft computing, fuzzy systems, and hybrid intelligent systems. Of particular value to both the contributors and the readership are the short publication timeframe and the world-wide distribution, which enable both wide and rapid dissemination of research output.

The books of this series are submitted to indexing to Web of Science, EI-Compendex, DBLP, SCOPUS, Google Scholar and Springerlink.

More information about this series at http://www.springer.com/series/7092

Stefka Fidanova

Editor

Recent Advances in Computational Optimization

Results of the Workshop on "Computational Optimization" and "Numerical Search and Optimization" 2018

 Springer

Editor
Stefka Fidanova
Parallel Algorithms
Institute of Information and Communication
Technologies, Bulgarian Academy
of Sciences
Sofia, Bulgaria

ISSN 1860-949X ISSN 1860-9503 (electronic)
Studies in Computational Intelligence
ISBN 978-3-030-22725-8 ISBN 978-3-030-22723-4 (eBook)
https://doi.org/10.1007/978-3-030-22723-4

This Springer imprint is published by the registered company Springer Nature Switzerland AG
The registered company address is: Gewerbestrasse 11, 6330 Cham, Switzerland

Organization

Workshop on Computational Optimization (WCO 2018) is organized in the framework of Federated Conference on Computer Science and Information Systems FedCSIS—2018 and Numerical Search and Optimization part of the Numerical Methods and Application Conference–2018.

Conference Co-chairs

For WCO
Stefka Fidanova, IICT-BAS (Bulgaria)
Antonio Mucherino, IRISA (Rennes, France)
Daniela Zaharie, West University of Timisoara (Romania)
For NSO
Stefka Fidanova, IICT-BAS (Bulgaria)
Gabriel Luque, University of Malaga (Spain)
Kalin Penev, Southampton Solent University (UK)

Program Committee

Bonates, Tibérius, Universidade Federal do Ceará, Brazil
Breaban, Mihaela, University of Iasi, Romania
Chira, Camelia, Technical University of Cluj-Napoca, Romania
Gonçalves, Douglas, Universidade Federal de Santa Catarina, Brazil
Hosobe, Hiroshi, National Institute of Informatics, Japan
Iiduka, Hideaki, Kyushu Institute of Technology, Japan
Lavor, Carlile, IMECC-UNICAMP, Campinas, Brazil
Marinov, Pencho, Bulgarian Academy of Science, Bulgaria
Micota, Flavia, West University Timisoara, Romania

Preface

Many real-world problems arising in engineering, economics, medicine, and other domains can be formulated as optimization tasks. Every day we solve optimization problems. Optimization occurs in the minimizing time and cost or the maximization of the profit, quality, and efficiency. Such problems are frequently characterized by non-convex, non-differentiable, discontinuous, noisy or dynamic objective functions, and constraints which ask for adequate computational methods.

This volume is a result of very vivid and fruitful discussions held during the Workshop on Computational Optimization and Workshop on Numerical Search and Optimization. The participants have agreed that the relevance of the conference topic and quality of the contributions have clearly suggested that a more comprehensive collection of extended contributions devoted to the area would be very welcome and would certainly contribute to a wider exposure and proliferation of the field and ideas.

The volume includes important real problems like modeling of physical processes, wildfire and natural hazards modeling, modeling metal nanostructures, workforce planning, wireless network topology, parameter settings for controlling different processes, extracting elements from video clips, and management of cloud computing environment. Some of them can be solved by applying traditional numerical methods, but others need huge amount of computational resources. Therefore for them are more appropriate to develop an algorithm based on some metaheuristic methods like evolutionary computation, ant colony optimization, particle swarm optimization, constraint programming, etc.

Sofia, Bulgaria
March 2019

Stefka Fidanova
Co-Chair
WCO'2018

Contents

Developing a Method for Measuring the Failover Times of First Hop Redundancy Within Video Networks 1
Paul Bourne, Neville Palmer and Jan Skrabala

Desktop Application Developed by Open Source Tools for Optimizations in Cases of Natural Hazards and Field Response 17
Nina Dobrinkova and Stefan Stefanov

Data Optimizations on Kresna Fire (2017) as Inputs for WFA Simulations 31
Nina Dobrinkova

Solving Sorting of Rolling Stock Problems Utilizing Pseudochain Structures in Graphs 45
Jens Dörpinghaus and Rainer Schrader

InterCriteria Analysis of Different Hybrid Ant Colony Optimization Algorithms for Workforce Planning 61
Stefka Fidanova, Olympia Roeva, Gabriel Luque and Marcin Paprzycki

Different InterCriteria Analysis of Variants of ACO algorithm for Wireless Sensor Network Positioning 83
Olympia Roeva and Stefka Fidanova

Geometric Versus Arithmetic Difference and H^1-Seminorm Versus L^2-Norm in Function Approximation 105
Stefan M. Filipov, Ivan D. Gospodinov, Atanas V. Atanassov and Jordanka A. Angelova

Ellipsoidal Estimates of Reachable Sets for Nonlinear Control Systems with Bilinear Uncertainty 121
Tatiana F. Filippova

Structural Instability of Gold and Bimetallic Nanowires Using Monte Carlo Simulation 133
Vladimir Myasnichenko, Nickolay Sdobnyakov, Leoneed Kirilov, Rossen Mikhov and Stefka Fidanova

Manipulating Two-Dimensional Animations by Dynamical Distance Geometry 147
Antonio Mucherino

Index Matrices as a Cost Optimization Tool of Resource Provisioning in Uncertain Cloud Computing Environment 155
Velichka Traneva, Stoyan Tranev and Vassia Atanassova

Three-Dimensional Interval-Valued Intuitionistic Fuzzy Appointment Model 181
Velichka Traneva, Vassia Atanassova and Stoyan Tranev

Logical Connectives Used in the First Bulgarian School Books in Mathematics 201
Velislava Stoykova

An Improved "Walk on Equations" Monte Carlo Algorithm for Linear Algebraic Systems 215
Venelin Todorov, Nikolay Ikonomov, Stoyan Apostolov, Ivan Dimov, Rayna Georgieva and Yuri Dimitrov

Author Index 237

Developing a Method for Measuring the Failover Times of First Hop Redundancy Within Video Networks

Paul Bourne🄳, Neville Palmer and Jan Skrabala

Abstract IP protocols have been used to distribute compressed media over private and public networks for a number of years. Recently the broadcast sector has started to adopt IP technologies to transport real time media within and between their facilities during production. However, the high bitrate of uncompressed media and its sensitivity to latency and timing variations requires careful design of the network in order to maintain quality of service. Connectionless protocols are commonly used, which means that packet loss is of particular concern and redundant paths must be provisioned with mechanisms to switch between them. This project develops and critically analyses a method for measuring the effectiveness of first hop redundancy protocols for broadcast video production networks. The aim extends previous work [1] to recommend particular configurations to optimise networks and to provide a method that broadcast engineers can use to verify performance. Cisco's HSRP is recommended with static routes configured for redundant paths. It is recommended that the network is tested using a synthetic RTP stream with a low complexity packet sniffer and NICs with hardware timestamps. Further work is identified including ways to improve the accuracy of the results and to consider the impact of more complex networks.

Keywords FHRP · HSRP · GLBP · VRRP · Network redundancy · Gateway protocol · Failover · Network performance measurement · Video network

P. Bourne (✉) · N. Palmer
Solent University, East Park Terrace, Southampton, Hampshire SO14 0YN, UK
e-mail: paul.bourne@solent.ac.uk

N. Palmer
e-mail: neville.palmer@solent.ac.uk

J. Skrabala
Vostron Ltd, 32B Castle Way, Southampton, Hampshire SO14 2AW, UK
e-mail: jan@vostron.com

© Springer Nature Switzerland AG 2020
S. Fidanova (ed.), *Recent Advances in Computational Optimization*,
Studies in Computational Intelligence 838,
https://doi.org/10.1007/978-3-030-22723-4_1

1

1   Introduction

The broadcast industry is currently undergoing a step change as it replaces bespoke network infrastructures based on the Serial Digital Interface (SDI) with commodity enterprise equipment using Internet Protocol (IP). Consumers in the UK are supplementing traditional broadcast channels with online services; TVB Europe's recent outlook report [2] states that viewers prefer streaming services over offline downloads and continues that such services are predicted to grow by 9.1% each year over the next five years. Meyer and Francis [3] suggest that the release of a number of IP related standards throughout 2017, as well as native IP products, has also increased broadcasters' confidence in IP technologies to the point that they are now investing heavily in IP infrastructures for playout and distribution. Broadcast services provider, Timeline Television, recently unveiled the first all IP outside broadcast truck [4]. The result is a desire for IP based infrastructures throughout the broadcast chain from capture through to delivery.

There are many advantages in terms of cost and flexibility but maintaining Quality of Service (QoS) for signals that are both time-sensitive and of high bandwidth can be challenging. Broadcasters are used to fixed bandwidth links and bespoke health monitoring mechanisms. In contrast network infrastructures have evolved with different requirements and have therefore developed significantly different techniques and monitoring tools. In recent years there have been significant efforts to help the sectors to converge, led by organisations such as the Broadcast Bridge and Alliance for IP Media Solutions [5, 6].

Higher level applications are often used to monitor network services and create reliable circuits using technologies such as Software Defined Networks (SDN) and Mulit Protocol Label Switching (MPLS). These tools are usually provided by third-party vendors based on proprietary intellectual property and produce a high degree of abstraction [7–9]. There are standard redundancy protocols available that can be configured to respond to changes in network performance. However, monitoring QoS in IP networks has proved to be difficult for broadcasters as the metrics built into enterprise equipment tend to be geared towards the operational requirements of generic data networks. The levels of abstraction and limited detail available may leave broadcast engineers wary of the technologies and either less likely to adopt them or to wildly overprovision the networks. The broadcast industry recognises the need for network engineers who understand the requirements of media signal flows and are starting to understand that a detailed understanding of networks is required to configure effective production LANs [10–12]. A review of current literature provides limited guidance on the relative performance of redundancy protocols and there are no procedures for measuring the performance of high availability networks.

This project aims to evaluate the effectiveness of existing network redundancy protocols and their suitability for broadcast video networks with a focus on production networks. The project will explore and evaluate methods of monitoring network performance and recommend a test procedure to allow engineers to verify and optimise their configurations.

2 Background

IP networks are becoming increasing important throughout the broadcast video work-flow from production through to distribution. Value can be added to the signal chain via the flexibility of dynamically routed signals and format agnostic transport protocols. There has been significant uptake of services distributed using IP technologies such as Over the Top (OTT) and Internet Protocol Television (IPTV) [13, 14]. OTT services such as Netflix and BBC iPlayer are delivered over the public Internet without the need for proprietary devices whereas IPTV services such as Virgin Media use managed networks. Due to the nature of the networks OTT is usually delivered using Hyper Text Transport Protocol (HTTP) using the connection-orientated Transmission Control Protocol (TCP). With the use of a buffer, this provides benefits such as retransmission but at the expense of latency. IPTV or real-time signals within a production environment are more likely to use Realtime Transport Protocol (RTP) over the connectionless User Datagram Protocol (UDP). RTP is used by many media-centric protocols as a transport mechanism to provide sequence numbers and timestamps with minimal overhead. The information provided within the headers may be used by high level applications to improve QoS and further feedback is usually exchanged between participants using the RTP Control Protocol (RTCP) [15]. The small receive buffer and lack of a retransmission mechanism means that RTP is vulnerable to packet-loss and necessitates a rapid failover mechanism to maintain resilience using a secondary path.

First Hop Redundancy Protocols (FHRP) provide an essential tool for increasing availability in critical switched IP networks. They provide a mechanism for fast failover to the next hop from a primary path to a secondary path within a group of backup routers. The process is faster than waiting for spanning tree or dynamic routing protocols to converge on a new path due to the limited scope and pre-configuration of the FHRP. Essentially two or more routers are able to share the default gateway at OSI layer 3, which provides an alternative route or may even be used for rudimentary load balancing. This does necessitate the use of multilayer switches but can be implemented at the Access Layer or Distribution Layer [16, 17]. YanHua and WeiZhe [18] have shown that such protocols are suitable for use within cable television IP networks when combined with device redundancy, although they caution against diminishing returns as the network complexity increases.

Cisco has developed two major proprietary protocols Hot Standby Routing Protocol (HSRP) and Gateway Load Balancing Protocol (GLBP). Another common protocol is the Virtual Router Redundancy Protocol (VRRP), which is available as an open IEFT standard RFC5798. It is similar to HSRP in operation but not compatible [19, 20]. Other equipment vendors also have proprietary protocols; Juniper Networks has NetScreen Redundancy Protocol (NSRP), Avaya has Routed Split Multi-link Trunking (R-SMLT) and Extreme Networks has Extreme Standby Routing Protocol (ESRP) [21–24]. Broadcast installations usually combine best-of-breed equipment and are generally multivendor environments built on open standards. However Cisco have a dominant position in the switching and routing market with over 50% of the worldwide market share in 2016 [25]. As such this investigation will focus on HSRP, GLBP and VRRP. Common Address Redundancy Protocol (CARP) was considered but relies on the Berkeley Software Distribution so is not practical on many platforms [26].

HSRP is configured for an interface using the standby command and allows the user to configure a virtual gateway for the connected hosts to use. Priorities are specified such that an active router is allocated with one or more standby routers sharing the virtual address with the active router as shown by Fig. 1. Packets are forwarded based on an IP/MAC address pair and standby routers monitor the status of the active router to promote a backup router in the case of a link failure on the active path. Tracking objects can be used to monitor interfaces or Service Level Agreement (SLA) tracking can monitor connectivity beyond the first hop. Either method can update the router priorities to determine the active path. Different priorities can also be assigned to different Virtual Local Area Networks (VLAN) to implement basic load balancing, although this may become unwieldy on large networks [17, 27].

GLBP uses multiple gateways simultaneously, which enables more effective load balancing and therefore uses all the bandwidth within the topology. Routers within a GLBP group may be the Active Virtual Gateway (AVG), an Active Virtual Forwarder (AVF) or the Standby Virtual Gateway (SVG). The AVG assigns virtual MAC address to the other group members. Up to four AVFs, including the AVG, are able to forward packets and the SVG is ready to take over from the AVG based on a similar priority system to HSRP with decrements based on tracking objects. GLBP is implemented on an interface using the glbp command and load balancing can be achieved within the group by assigning packets to the MAC addresses of the AVFs via an equal round-robin, by weighting certain paths or based on the host [17, 28].

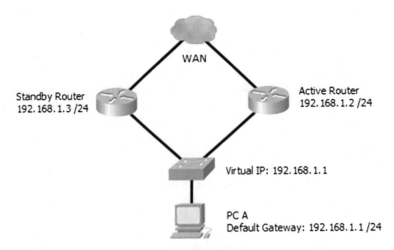

Fig. 1 A typical HSRP configuration with the host using a common virtual gateway IP address

VRRP is very similar to HSRP in that is uses a single virtual gateway that is shared between a master and one or more backup router. It is implemented on the interface using the vrrp command and supports object tracking to determine failures. Fewer IP addresses may be used by VRRP than HSRP as the physical IP address for the master router may also be used as the virtual IP address [29].

Several articles have been published that outline the configuration options for HSRP, GLBP and VRRP [20, 30–32]. These discuss how the options affect the underlying algorithm but there appears to be little guidance on how to optimise the parameters or how the common protocols compare under similar conditions. Ibrahimi et al. [32] and Rahman et al. [33] demonstrate a basic method, using continuous Internet Control Message Protocol (ICMP) echo requests to show the duration of a link failure. Pavlik et al. [17] show a more accurate method to determine the interval between missing and restored replies using timestamps from a packet sniffer. None of the studies investigate the time taken to restore a link after the primary link recovers.

QoS describes the technical performance of a network, which can usually be objectively quantified and measured at specific points within the network topology. Common parameters include latency, jitter and packet loss but bandwidth and packet reordering are also important within video networks due to the temporal sensitivity and inter-packet dependencies of media flows [34, 35]. Latency is the time it takes for packets to reach their destination whereas jitter is the variance in the inter-packet latency. Latency is often acceptable when there is minimal jitter or packet loss, although it can be a problematic in live broadcasting when disparate sources are combined such as when conducting interviews. Packet loss tends to be caused by excessive latency or jitter that causes buffers to underflow and tends to result in audio or video frames freezing or degrading. Receive buffers can reduce jitter and therefore packet loss but at the expense of latency [36]. Bandwidth is of concern within production facilities due to the high data rates of video at 2.97 Gbps for a single uncompressed High Definition (HD) stream. Gharai, Perkins and Lehman [37] have noted that in video networks, packet reordering should be treated on an equal footing with packet loss and the European Broadcast Union (EBU) recommends that a receiver should not have to accommodate packets out of order by more than 10 places. Packet reordering is often caused during redundancy switches. Cisco recommends that for video networks latency should be less than 300 ms, jitter less than 50 ms and packet loss of less than 0.5% [38] although production environments may have to work to stricter limits for control and monitoring with latency as low as 10 ms and virtually no packet loss [39, 40].

3 Test Method

How to measure the effectiveness of FHRPs is one of the major objectives of the research and a clear and accessible method should be one of the outputs. Previous studies of the quality of video over IP have made a distinction between the QoS provision from the network and the QoS provision from the media application [36,

41]. Studies at the application layer have tended to focus on comparing image quality by detecting artefacts such as blocking and blurring or calculating Peak Signal to Noise Ratio (PSNR) using bespoke measurement applications. These are very dependent on the configuration of the codec and often require reliable side-channels to make mathematical comparisons with a reference stream. This study will instead focus on the QoS of the network, which Tao, Apostolopoulos and Guerin [41] state is largely driven by packet loss, delay and jitter. This may be achieved by generating and injecting a stream of timestamped packets into a test network to be captured and analysed at a receiver. Previous studies into FHRPs demonstrate a method to test availability using a continuous stream of Internet Control Message Protocol (ICMP) echo requests to determine the duration of a link failure [32, 33]. Alternatively a packet sniffer can be used to observe the interval between missing echo relies as demonstrated by Pavlik et al. [17]. Essentially packet drop and the unavailability interval provide the same information as the packet drop is a function of the interval and bitrate.

There are several network simulation packages that may be used to rapidly analyse the behaviour of different configurations. Common simulators include Cisco's Packet Tracer, open source application GNS3 and Riverbed Modeler. For these to produce accurate results, they require detailed implementations of the software and protocols running on the network devices as well as accurate models of the hardware performance. The level of difficulty required to accurately reproduce the test environment is not necessary where real equipment is available to provide accurate real time results.

A testbed was created to emulate a typical spine-and-leaf network architecture as recommended for high available networks. The topology is shown by Fig. 2.

The source had an edge 'customer' router, which connected to their Local Area Net-work (LAN). This had redundant links to an external network via primary and secondary routers, which were connected to the 'main' router at the destination on a WAN. The switch simplified the configuration of the customer router by removing the need for two interfaces on the same subnet. The primary link utilised a Gigabit Ethernet connection whereas the secondary was only Fast Ethernet. Secondary links are often metered in practice so load balancing was not implemented. To simulate a link failure, the interface G0/0 was shut down on the primary router with a tracking object to promote the secondary link based on the line-protocol state. The FHRPs were configured to decrement the priority of the primary router below that of the secondary router if the line protocol went down; this causes the secondary router to pre-empt the primary router and traffic would be rerouted. An alternative method would be to check the reachability of the loopback interface on the destination network using ICMP requests to decrement the priority upon failure. This would be a more meaningful detection method within a real network but the non-deterministic nature of packet generation and propagation may distort the results, which should be focused on the responsiveness of the redundancy protocol itself.

All routes were statically defined within the routers to prevent the dynamic routing protocols from interfering with the results. The FHRPs were configured on the primary and secondary routers with a default static route on the customer router

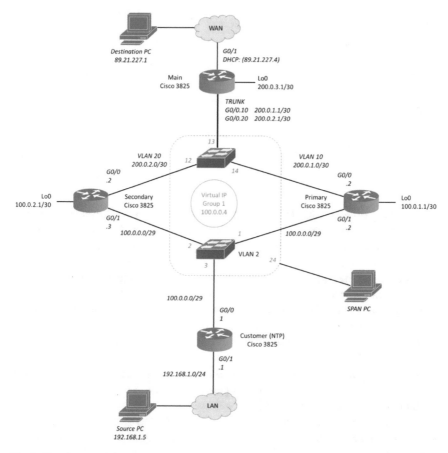

Fig. 2 Topology for FHRP experiment

pointing to the virtual IP address. A default and more rapid configuration was tested for each FHRP using the parameters shown by Table 1. By default HSRP sets the time between hello messages to 3 s and the hold time to be 10 s. The hold time is the interval after which the active router is declared to be down once hello messages are not acknowledged. The hold timer must be greater than the hello timer—usually at least three—to avoid the active router being declared down between hello messages [42, 43]. GLBP uses the same system of hello and hold timers with identical default and minimum values [44]. HSRP and GLBP timers can be set with millisecond granularity but a shorter timer increases the amount of overhead traffic and can cause the system to behave erratically as it becomes too sensitive [34]. For the rapid test the hello timer will be set to 1 s and the hold timer to 3 s; this will allow the effect of smaller timers to be observed without creating instability. VRRP advertisements are sent every second by default with a failover delay calculated by three times the advertisement interval plus the router's skew time. The skew time is based on the

Table 1 Active variables for FHRP experiment

	Hello interval (s)	Hold timer (s)	Pre-emption delay (s)	Priority (backup/master)
HSRP D	3	10	0	–
HSRP R	1	3	0	–
GLBP D	3	10	30	–
GLBP R	1	3	3	–
VRRP D	1	–	0	100/110
VRRP R	1	–	0	253/254

inverse of the priority. The standard doesn't include shorter advertisement intervals but Cisco has implemented this in their IOS down to 20 ms [30, 45]. The priorities were configured as high as possible for the rapid test to reduce the skew time. It should be noted that pre-emption delay should normally be configured to allow for the boot time of the equipment [34]. Each configuration was tested ten times.

Two methods were employed for measuring the interval between the interface changing state and the FHRP responding. A third method that would track missing ICMP echo replies was considered and trialed during earlier studies [1]; the results from this would be more comparable to existing studies but the flow was found to be too simplistic and imprecise to represent media streams.

Method 1: The packet loss was measured using a test stream and packet sniffer. This relies on the accuracy of the packet generator as well as the timestamps within the captured stream. VLC Media Player was used to generate a stream of colour bars and tone from an MP4 file. netsniff-ng was used to capture traffic for analysis using Wireshark's RTP dissector. The interval between missing sequence numbers could be used to determine the packet loss during failover or restoration. The jitter on the captured packets was also measured to determine the inter-packet arrival variance and therefore the precision to which the time interval could be reliably calculated.

Method 2: Timestamped debug messages on the routers were used to report the interface failure and pre-emption of a new router. NTP was used to synchronise the internal clocks so that the timestamps from different devices could be compared. The accuracy of the log timestamps was investigated by sending 1000 PING requests from the source PC to the primary router with 100 ms interval. The SPAN feature on the primary router was used to copy packets from the G0/1 interface to a netsniff-ng packet capture on the SPAN PC. Logging was configured on the primary router to millisecond accuracy using an Access Control List (ACL) to also log the arrival of the ICMP echo requests. The timestamps on the packet capture were compared to the logging timestamps to indicate any discrepancy with the logging process. The jitter inherent on the network was also measured using the Iperf utility.

4 Results

Of the 1000 PING requests sent, the maximum deviation from the intended 100 ms interval was ± 4 ms within the raw Cisco timestamps. The PING utility and source interface will introduce jitter so the figures were adjusted based on the inter-packet arrival intervals from the SPAN capture. This reduced the error slightly to +4 ms and −3 ms. In total 31 packets were logged as having a different arrival time to the SPAN data with an average absolute deviation of 2 ms. The round trip time reported by the PING utility ranged from 1 to 2 ms with an average of 1.88 ms.

After running 10 cycles of the jitter measurements, the maximum jitter reported was 570 μs, the minimum 26 μs and the mean average result was 82 μs with a standard deviation of just 0.1.

The results from the experiment suggest that Cisco's logging timestamps have the potential to be accurate to within +4/−3 ms. The jitter added from the test equipment was much lower than this, which suggests that the figures are reliable. This is enough precision to compare FHRP but Accedian [46] suggest that inaccuracies in timestamps approaching even a tenth of a millisecond are unsuitable for measuring high performance SLA metrics. One concern is that the load on the router was extremely low during this experiment as there was no significant traffic being routed or other events for the router to log. As such this indicates the best accuracy of the logging timestamp.

The minimum, maximum and mean average time have been presented for the default and rapid configurations of each redundancy protocol. The failover times are summarised in Table 2 and the restoration times in Table 3. The average jitter calculated from the stream captures is included to provide an indication of the system stability—jitter data has not been presented for the rapid configuration of VRRP as it was consistently reported to be zero. The mean absolute differences between the results from the two methods are shown as absolute value and percentages.

The mean failover and restoration delays as well as the mean number of packets dropped are shown as radar plots for the default and rapid configuration in Fig. 3 and Fig. 4 respectively. The metrics have been scaled to produce a plot whereby the outer edge indicates the worst performance and the center indicates ideal performance.

It can be seen from Tables 2 and 3 that the two test methods produce similar results for VRRP and GLBP but there is significant disparity for HSRP—the captures suggest that the protocol is 83–88% faster than the logs report. Inspection of the packet captures verifies that this figure is correct and suggests that the logging process introduces a delay with this protocol. The results correlate well with the studies completed by Pavlik et al. [17] although it is interesting to note that all of the protocols were faster than in their study. This is likely to be due to the different topologies used and will be particularly affected by the use of static routes in this study. Previous studies have not analysed the restoration delays, which were also found to be fastest with HSRP closely followed by VRRP.

The packet loss showed some particularly interesting results. The rapid failover of HSRP led to the fewest losses whilst GLBP resulted in the most. Surprisingly

Table 2 FHRP failover delays in seconds from packet capture and cisco logs

| | PCap RTP sequence errors | | | | |
	Min	Max	Mean	Jitter	Dropped
HSRP D	0.325	2.576	1.527	0.069	29
HSRP R	0.313	1.026	0.655	0.069	9
VRRP D	3.090	3.937	3.461	0.071	76
VRRP R	2.244	2.874	2.602	–	79
GLBP D	30.705	38.928	33.425	0.069	670
GLBP R	3.913	4.894	4.332	0.071	678
	Cisco logs			Mean absolute difference (%)	
	Min	Max	Mean		
HSRP D	11.188	13.344	12.280	10.753	87.56
HSRP R	0.644	3.856	3.274	2.713	82.87
VRRP D	2.784	3.552	3.206	0.307	9.57
VRRP R	2.192	2.900	2.548	0.581	22.81
GLBP D	30.464	34.984	32.337	1.107	3.42
GLBP R	3.652	4.580	4.047	0.285	7.04

Table 3 FHRP restoration delays in seconds from packet capture and cisco logs

| | PCap RTP sequence errors | | | | |
	Min	Max	Mean	Jitter	Dropped
HSRP D	0.293	2.692	1.484	0.069	0
HSRP R	0.417	1.009	0.725	0.069	0
VRRP D	2.360	3.535	3.132	0.071	574
VRRP R	2.205	2.878	2.508	–	0
GLBP D	31.200	46.519	35.533	0.069	0
GLBP R	2.993	3.358	3.196	0.071	0
	Cisco logs			Mean absolute difference (%)	
	Min	Max	Mean		
HSRP D	0.368	2.660	1.436	0.083	5.77
HSRP R	0.156	0.944	0.633	0.081	12.83
VRRP D	2.840	3.532	3.166	0.168	5.31
VRRP R	2.288	2.924	2.698	0.239	8.88
GLBP D	31.108	46.512	34.654	0.874	2.52
GLBP R	3.100	3.196	3.128	0.103	3.28

Fig. 3 A comparison of protocol performance with default configurations

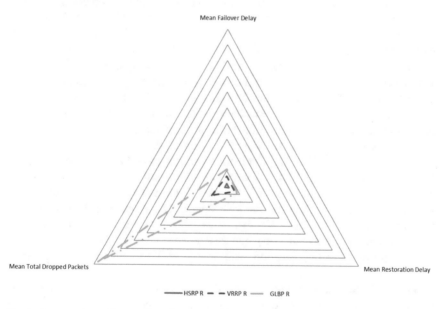

Fig. 4 A comparison of protocol performance with rapid configurations

GLBP resulted in slightly more loss with its rapid configuration even though the link initially came back up more quickly. VRRP consistently dropped a significant number of packets during restoration when configured with the default parameters—no other configuration exhibited this behaviour including the rapid configuration of VRRP. This is highly undesirable as it will result in an additional video glitch in the event of a failover.

5 Conclusions

The initial tests logging ICMP echo requests suggested that timestamps could be accurate to within ± 4 ms and that the jitter from low-level packet generation tools would have minimal influence on the results. However, the disparity between the logs and packet captures show that the timestamps from logs cannot be relied upon for more complex protocols—this could be due to the interface between the protocol and the logging procedure or due to the priority that is assigned to logging the protocol. It is recommended that it is more accurate to capture an RTP stream to test the failover delay.

Originally packEth was going to be used to generate an RTP stream. The threading used to generate the traffic is reported to be stable to microsecond resolution [47] and Srivastava et al. [48, 49] found the tool to be more suitable than alternatives for testing high capacity links. However this tool doesn't produce RTCP messages, which caused the routers to drop the RTP stream. Instead VLC Media Player was used, which had to encode real video frames rather than generating a synthetic stream. The captures showed a throughput of approximately 20–30 packets per second, which didn't give much resolution and the delta between consecutive packets varied from approximately 0.04 to 250 ms, which is quite bursty. The resolution could be improved by increasing the data-rate at the source but the burstiness is likely to be harder to solve. The jitter in the network was fairly consistent throughout the experiment at around 70 ms. A video network would be expected to have jitter below 30 ms to maintain QoS [50]. The high figure is likely to be due to using a software application to generate the RTP stream. The worst case jitter originating from the source and capture NICs themselves was measured to be just 0.57 ms. A tool is needed to synthetically generate RTP streams with RTCP messaging so that a high throughput can be generated with low jitter. This could be an extension to packEth or a separate utility.

The results themselves suggest that HSRP is the most effective protocol and that the rapid configuration improves its performance. This protocol is Cisco proprietary so VRRP would have to be used within a multivendor environment with the configuration customised to avoid the packet loss observed using the default parameters. This work could be extended by reducing the configured timers further to find the limits of the protocol. As broadcasters are likely to use larger systems with best of breed equipment, the effects of multiple standby devices and mixed-vendor equipment should also be investigated.

References

1. Bourne, P., et al.: Quantifying the effectiveness of first-hop redundancy protocols in IP networks. In: Numerical Methods and Applications, NMA 2018. Lecture Notes in Computer Science, vol 11189. Springer, Cham (2019)
2. Ramsey, C.: 2018 Mediatech Outlook. Newbay, s.l. (2018)
3. IP in 2018—A Growth in Market Confidence. https://www.tvtechnology.com/news/ip-in-2018-a-growth-in-market-confidence. Accessed 27 Jan 2019
4. Timeline Television to Showcase 4 K-HDR IP OB Truck on SAM's IBC 2017 Booth. https://www.thebroadcastbridge.com/content/entry/9387/timeline-television-to-showcase-its-ip-4k-hdr-ob-truck-on-sams-ibc-2017-boo. Accessed 21 Jan 2019
5. About—The Broadcast Bridge. https://www.thebroadcastbridge.com/about. Accessed 21 Jan 2019
6. Overview—AIMS Alliance. https://www.aimsalliance.org/overview/. Accessed 21 Jan 2019
7. Man, H., Li, Y.: Multi-stream video transport over MPLS networks. In: IEEE Workshop on Multimedia Signal Processing, pp. 384–387 (2002)
8. Aria Networks Shows Off Self-Healing Video Distribution Using SDN + AI Combo. http://www.aria-networks.com/news/aria-networks-self-healing-video-distribution-sdn-ai/. Accessed 21 Jan 2019
9. Demystifying SDN: It's an integral part of next-gen IP video transport. https://www.tvtechnology.com/expertise/demystifying-sdn. Accessed 21 Jan 2019
10. TV, Vendors Struggle To Keep Up With Tech. https://tvnewscheck.com/article/114036/tv-vendors-struggle-to-keep-up-with-tech/. Accessed 21 Jan 2019
11. What Broadcasters Can Learn From Telcos About IP. https://www.tvtechnology.com/broadcast-engineering/what-broadcasters-can-learn-from-telcos-about-ip. Accessed 21 Jan 2019
12. Whether It's Called IP or IT, Media Networks Are Different. https://www.thebroadcastbridge.com/content/entry/7062/whether-its-called-ip-or-it-media-networks-are-different. Accessed 21 Jan 2019
13. Quarter of European OTT platforms target cross-border audiences. https://www.broadbandtvnews.com/2017/11/15/quarter-of-european-ott-platforms-target-cross-border-audiences/. Accessed 21 Jan 2019
14. The revolutionary transition to an all-IP video production workflow. https://www.ibc.org/production/revolutionary-transition-to-all-ip-video-production-workflow/925.article. Accessed 21 Jan 2019
15. Cheng, Y., et al.: From QoS to QoE: a tutorial on video quality assessment. IEEE Commun. Surv. Tutor. **17**(20), 1126–1165 (2015)
16. CCNA Routing and Switching Practice and Study Guide: LAN Redundancy. http://www.ciscopress.com/articles/article.asp?p=2204384&seqNum=4. Accessed 21 Jan 2019
17. Pavlik, J., et al.: Gateway redundancy protocols. In: 15th IEEE International Symposium on Computational Intelligence and Informatics, pp. 459–464. IEEE, Budapest (2014)
18. YanHua, Z., WeiZhe, M.: The design of cable television IP access network based on hot standby router protocol. In: International Conference on Image Analysis and Signal Processing, pp. 1–4. IASP, Hangzhou (2012)
19. Virtual Router Redundancy Protocol (VRRP) Version 3 for IPv4 and IPv6. https://tools.ietf.org/html/rfc5798. Accessed 21 Jan 2019
20. Oppenheimer, P.: Top-Down Network Design, 3rd edn. Cisco Press, Indianapolis (2011)
21. OpenBSD PF—Firewall Redundancy (CARP and pfsync). https://www.openbsd.org/faq/pf/carp.html. Accessed 21 Jan 2019
22. NSRP overview. https://kb.juniper.net/InfoCenter/index?page=content&id=KB4263&cat=HA_REDUNDANCY&actp=LIST. Accessed 21 Jan 2019
23. Routed split multilink trunking. https://worldwide.espacenet.com/publicationDetails/biblio?FT=D&date=20050113&DB=&locale=en_EP&CC=US&NR=2005007951A1&KC=A1&ND=1. Accessed 21 Jan 2019

24. ExtremeWare XOS Concepts Guide Software Version 11.3, p. 445. https://www.manualslib. com/manual/511330/Extreme-Networks-Extremeware-Xos-Guide.html?page=1#manual. Accessed 4 July 2018

25. Switching market stays steady as Cisco's domination continues. https://www.cbronline. com/data-centre/switching-market-stays-steady-as-ciscos-domination-continues-4993282/. Accessed 21 Jan 2019

26. Chapter 28. Introduction to the Common Address Redundancy Protocol (CARP). https://www. netbsd.org/docs/guide/en/chap-carp.html. Accessed 21 Jan 2019

27. Hot Standby Router Protocol Features and Functionality. https://www.cisco.com/c/en/us/ support/docs/ip/hot-standby-router-protocol-hsrp/9281-3.html. Accessed 21 Jan 2019

28. Cisco: First Hop Redundancy Protocols Configuration Guide. Cisco IOS Release 12.2SX. Cisco, San Jose (2011)

29. VRRP Overview. https://www.juniper.net/documentation/en_US/junose15.1/topics/concept/ vrrp-overview.html. Accessed 21 Jan 2019

30. First Hop Redundancy Protocols Configuration Guide, Cisco IOS Release 15SY. https://www. cisco.com/c/en/us/td/docs/ios-xml/ios/ipapp_fhrp/configuration/15-sy/fhp-15-sy-book/fhp-vrrp.html. Accessed 21 Jan 2019

31. Configuring VRRP. https://www.juniper.net/documentation/en_US/junos/topics/example/ vrrp-configuring-example.html. Accessed 21 Jan 2019

32. Ibrahimi, M., et al.: Deploy redundancy of Internet using first Hop redundancy protocol and monitoring it using IP service level agreements. Int. J. Eng. Sci. Comput. 7(10), 15320–15322. Pearl Media Publications, Bangalore (2017)

33. Rahman, Z., et al.: Performance evaluation of first HOP redundancy protocols (HSRP, VRRP & GLBP). J. Appl. Environ. Biol. Sci. 7(3), 268–278. Textroad, Egypt (2017)

34. Froom, R., Sivasubramanian, B., Frahim, E.: Implementing Cisco Switched Networks. Cisco, Indianapolis (2010)

35. Arthur, C., et al.: The effects of packet reordering in a wireless multimedia environment. In: 1st International Symposium on Wireless Communication Systems, pp. 453–457 (2004)

36. Cheng, Y., Wu, K., Zhang, Q.: From QoS to QoE: a tutorial on video quality assessment. IEEE Commun. Surv. Tutor. 17(20), 1126–1165 (2015)

37. Gharai, L., Perkins, C., Lehman, T.: Packet reordering, high speed networks and transport protocol performance. In: 13th International Conference on Computer Communications and Networks, 1401591, pp. 73–78 (2004)

38. Video Quality of Service (QOS) Tutorial. https://www.cisco.com/c/en/us/support/docs/quality-of-service-qos/qos-video/212134-Video-Quality-of-Service-QOS-Tutorial.html. Accessed 21 Jan 2019

39. Understanding—and Reducing—Latency in Video Compression Systems. http://www. cast-inc.com/blog/white-paper-understanding-and-reducing-latency-in-video-compression-systems. Accessed 21 Jan 2019

40. IP Networks for Broadcaster Applications. https://www.researchgate.net/publication/ 308764506_IP_Networks_for_Broadcaster_Applications. Accessed 21 Jan 2019

41. Tao, S., Apostolopoulos, J., Guerin, R.: Real-time monitoring of video quality in IP networks. IEEE/ACM Trans. Netw. 16(5), 1052–1065 (2008)

42. Hot Standby Router Protocol (HSRP): Frequently Asked Questions. https://www.cisco.com/c/ en/us/support/docs/ip/hot-standby-router-protocol-hsrp/9281-3.html. Accessed 21 Jan 2019

43. Standby-hold-timer. https://www.ibm.com/support/knowledgecenter/SS9H2Y_7.5.0/com. ibm.dp.doc/standby-hold-timer_interface.html. Accessed 21 Jan 2019

44. Timers (GLBP). https://www.cisco.com/c/m/en_us/techdoc/dc/reference/cli/n5k/commands/ timers-glbp.html. Accessed 21 Jan 2019

45. VRRP failover-delay Overview. https://www.juniper.net/documentation/en_US/junos/topics/ concept/vrrp-failover-delay-overview.html. Accessed 21 Jan 2019

46. White Paper—One-Way Delay Measurement Techniques. https://www.accedian.com/wp-content/uploads/2015/05/One-WayDelayMeasurementTechniques-AccedianWhitePaper.pdf. Accessed 21 Jan 2019

47. Why-micro-seconds-are-not-enough. https://packeth.wordpress.com/2015/05/04/why-micro-seconds-are-not-enough/. Accessed 21 Jan 2019
48. Srivastava, S. et al.: Comparative study of various traffic generator tools. In: Recent Advances in Engineering and Computational Sciences (RAECS), pp. 1–6 (2014)
49. Srivastava, S. et al.: Evaluation of traffic generators over a 40 Gbps link. In: Asia-Pacific Conference on Computer Aided System Engineering (APCASE), pp. 43–47 (2014)
50. Implementing Quality of Service Over Cisco MPLS VPNs. http://www.ciscopress.com/articles/article.asp?p=471096&seqNum=6. Accessed 21 Jan 2019

Desktop Application Developed by Open Source Tools for Optimizations in Cases of Natural Hazards and Field Response

Nina Dobrinkova and Stefan Stefanov

Abstract In our paper we present a decision support desktop application that can provide the opportunity for faster response of firefighting and volunteer groups in cases of wildland fires or flood events. It is developed with free and open source software. The application is desktop based and is meant for operational room support. The desktop application has two modules. The first one is about wildland fires. The second is about flood events. The main goal is visualization of POI's (Points Of Interest) as logistic centers for water supplies and different tools needed on the field working teams. List of the tools is included in the POIs information. The application goal is to be calibrated on the field. The wildland fires module will be tested in municipalities of Zlatograd, Madan and Nedelino during the duration of eOutland project. The flood events module will be tested in Geghi reservoir, Armenia during the duration of ALTER project.

Keywords Open source software · Decision support application · QGIS

1 Introduction

The idea of the application is to visualize vulnerable objects and POI's. It can be used in operational room for support of firefighting and volunteer groups for field response in cases of wildland fires or flood events. The POI's are water supply locations and logistic centres that contain different firefighting tools for field work.

The application has two modules: Wildfire module [1] and flood module [2] for each of them we have different testing zones. The test zone about wildfire module is

N. Dobrinkova (✉) · S. Stefanov
Institute of Information and Communication Technologies-Bulgarian Academy
of Sciences Acad., Georgi Bonchev bl. 2, 1113 Sofia, Bulgaria
e-mail: ninabox2002@gmail.com

N. Dobrinkova · S. Stefanov
Center for National Security and Defense Research - Bulgarian Academy of Sciences, Sofia, Bulgaria
e-mail: stefans.stefanov303@gmail.com

© Springer Nature Switzerland AG 2020
S. Fidanova (ed.), *Recent Advances in Computational Optimization*,
Studies in Computational Intelligence 838,
https://doi.org/10.1007/978-3-030-22723-4_2

located on the territory of Zlatograd, Madan and Nedelino municipalities. The test zone about flood module is located in the Akhtala and Teghut areas of Lori Marz along the Shamlugh river. Including the Vorotan Cascade and its associated dams in the Syunik region, Kapan and Voghji river basin of Syunik area. Flood module [2] is visualizing the flood dangerous zones and the possibilities how the wave can propagate with time. Also the software give as option visualization of important buildings and infrastructure. The testing area is the Geghi reservoir located in Syunik, the southernmost province of Armenia. The needed geodatabases (infrastructure, buildings, administrative units, water objects, monitoring sites location, etc.) is provided from the American University of Armenia and Institute of Geological Sciences.

2   Main Features of the Desktop APP

The desktop application is fully developed with open source software [3]. The free sources of data and the open source software solutions that we are using for developing our desktop application give us a lot of opportunities and possibilities to build application that can be used in the operational rooms in support of volunteer and firefighter teams in cases of wildland fires and flood events. Our decision support application will have two modules: wildfire module and flood module [2].

The wildfire module [1] is mainly focused on the ability for providing better decision support to the groups when they are on the field. It has ability to visualize logistic centres and what equipment is located there. The base layer can be switched between different styles which depends on the needs of the users. Module can visualize the distance between two or more points on the map. The fire danger zones of previous fires can be also visualized. In the application is included quick link to the weather forecast maps. Zoom in, zoom out and printing options are available for the users. The tool has geolocation which is based on the network location (Fig. 1).

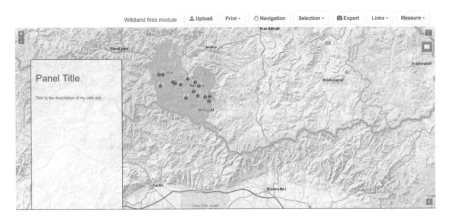

Fig. 1 Wildland fires module—main panel

Main features of the Wildland fires module:

- POI's feature: visualize the location of objects like water supplies and logistic centres and provide additional information about them as what kind of equipment is included inside;
- Base layers feature: gives the opportunity for base layer change, by inclusion of different map types as street map, earth view map, relief map etc. Users can chose the map style which fits best on their needs;
- Layer switch—give the opportunity of easy switching between predefined layers if is needed specific information;
- Measures feature: can measure the distance between different objects and may be used for field support;
- Weather forecast maps feature: this service is based on geolocation and visualize different data and maps about weather like: temperature, wind speed and direction etc.;
- Zoom feature: can be used for more detailed information for the objects;
- Print feature: can make digital maps printed on paper if this is necessary;
- Geolocation feature: based on network location, this feature visualizes the location of the device which is using the application;
- Selection feature: allows users to select part of the map for better performance;
- Extra layers import feature: users can import custom layers into the application if they need to visualise something more specific;
- Image export feature: can export map as an image file (Figs. 2 and 3).

The flood module [2] is focused to visualise how the high waves can spread in case of flood event. The module is showing the most vulnerable buildings in cases of flood hazard. This buildings type is: schools, kindergarten etc. It delivers information about the nearest and most threatened buildings. The tool purpose is to deliver geo data data to the groups on the field in fast and convenient format. This data can support them for better response and decision taking. This module is connected with The Alliance for Disaster Risk Reduction project with acronym: ALTER. The project focuses on establishment of public-private partnerships to understand and address flood risks that may stem from water and mining dam failures. Know-how, technologies and experience from the European Union are transferred to Armenia. The testing zones are the Akhtala and Teghut areas of Lori Marz along the Shamlugh river, the Vorotan

Fig. 2 Wildland fires module—base layers, measure and selection

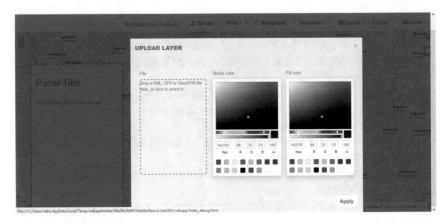

Fig. 3 Wildland fires module—custom layer upload

Fig. 4 Flood events module—main panel

Cascade and its associated dams in the Syunik region, and the Kapan and Voghji river basin of Syunik region (Fig. 4).

Main features of the flood events module:

- POI's feature: visualize the location of objects like water supplies and logistic centres and provide additional information about them. Special equipment which may be needed for the teams on the field;
- Base layers feature: give the opportunity for easy switch between any selected base layer such as street map, earth view map, relief map etc. Any user can chose the map style which fits the best on his/her needs;
- Layer switch—gives the opportunity of easy change between the predefined layers if it is needed specific information;
- Measures feature: can measure the distance between different objects on the field;

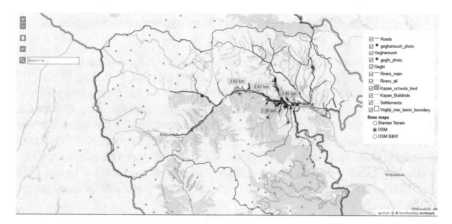

Fig. 5 Flood events module—base layers, measure and search

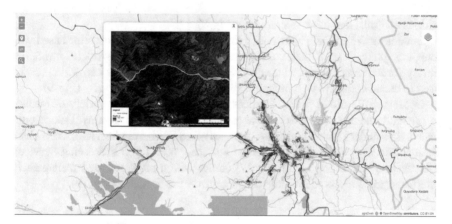

Fig. 6 Flood events module—orthophoto visualisation

– Zoom feature: can be used for more detailed information for the objects;
– Geolocation feature: is based on network location, this feature visualizes the location of the device which is using the application;
– Orthophoto visualization: this feature can be used as ground support for better visualization of the terrain and can deliver different data in help of groups on the field;
– Searching feature: it allows searching of different data like specific location, layer etc. (Figs. 5 and 6).

3   Evaluation of the Desktop Tool for Implementation Needs

In order to build our application for a Desktop usage we focused on free and open source software [3] for processing the data such as: Qgis [4], Web App Builder [5], Boundless WEBSDK [6], Geoserver [7], PostgreSQL [8] and OpenLayers [9].

QGIS [4] is a GIS client where users can visualize, manage, edit, analyze data and compose printable maps. It includes analytical functionality through integration with GRASS (Geographic Resources Analysis Support System), SAGA (System for Automated Geoscientific Analyses), Orfeo Toolbox, GDAL/OGR (Geospatial Data Abstraction Library) and many other algorithm providers. It runs on Linux, Unix, Mac OSX and Windows and supports numerous vector, raster and database formats and functionalities (Fig. 7).

Web App Builder [5] is a plugin for QGIS [4] that allows easy creation of web applications based on layers, map compositions and bookmarks, as configured within a QGIS [4] platform. The applications can also include additional web services, various controls, and other interactivity (Fig. 8).

Boundless WEBSDK [6] provides tools for easy-to-build JavaScript-based web mapping applications. It makes use of the JavaScript React framework (A JavaScript library for building user interfaces) to provide modular components, which can be used to create complete web mapping applications quickly and easily (Fig. 9).

Geoserver [7] allows users to process maps and data from a variety of formats to standard clients such as web browsers and desktop GIS programs. Data is published via standard based interfaces, such as WMS (Web Map Service), WFS (Web Feature Service), WCS (Web Coverage Service), WPS (Web Processing Service), Tile Caching and more. GeoServer [7] comes with a browser-based management interface and connects to multiple data sources at the back end (Fig. 10).

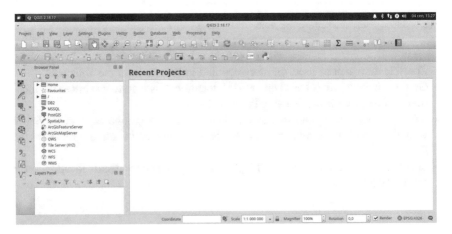

Fig. 7 QGIS GUI (Graphical User Interface)

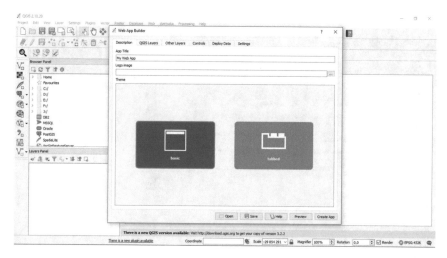

Fig. 8 Web App Builder plugin

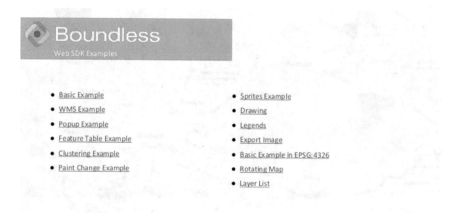

Fig. 9 Boundless WEBSDK GUI

PostgreSQL [8] is a powerful, open source object-relational database system that uses and extends the SQL language combined with many features that safely store and scale the most complicated data workloads also can store geospatial data. Post-greSQL [8] comes with many features aimed to help developers build applications, administrators to protect data integrity and build fault-tolerant environments, and help you manage your data no matter how big or small the dataset is. The system is available for Windows and Linux (Fig. 11).

Fig. 10 Geoserver admin page

Fig. 11 PostgreSQL GUI

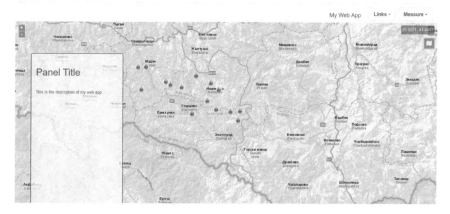

Fig. 12 OpenLayer example visualization

OpenLayers [9] makes it easy to put a dynamic map in any web page. It can display map tiles, vector data and markers loaded from any source. OpenLayers [9] has been developed to use geographic information of all kinds. It is completely free, Open Source JavaScript based. OpenLayers [9] supports GeoRSS, KML (Keyhole Markup Language), Geography Markup Language (GML), GeoJSON and map data from any source using OGC-standards as Web Map Service (WMS) or Web Feature Service (WFS) (Fig. 12).

4 How the System is Working

The Decision support application has two modules [1], ready-to-use Web GIS solution [3], where users can work with the pre-configured data layers.

The system is designed in such way that it can be used by anybody with average computer skills and a basic understanding of GIS. No special system administration and programming skills are required.

The basic tools are:
1. Wildfire module (Fig. 13):

- This module provide the ability to work with layers, which can be switched easily from the layer icon which opens the list with available layers and is placed on the top right part of the screen or the base layers, which can be changed depending of the user's needs;
- Zoom in and zoom out: can be used with mouse scroll or with the icons placed on the top left side of the screen;
- Geolocation: if the user need to locate himself, in this case can be used the icon placed on the top left side of the screen just under the zoom in and zoom out buttons;

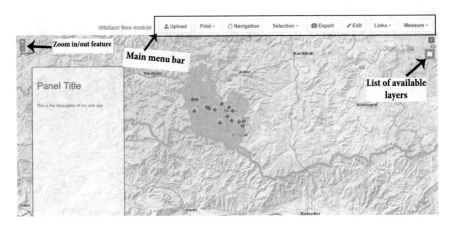

Fig. 13 Wildland fires module

– Measures: if the user need to measure distances or take measures of areas they can use the measure tool by clicking on the measure button placed in the menu bar. After user activates this feature by clicking on the map, he/she can measure the distance between 2 or more points or take measures of the area for deactivating this feature. If the user need to change measurement, than it has to be selected remove measures button;

– Selection feature: with this option any user can draw rectangles on map for more efficient work with data. It can be activated by clicking on the button Selection placed in the menu bar;

– Extra layers import feature: the users can import custom layers into the application if they need to visualise something more specific this feature can be activated by clicking on the button Upload placed in the menu bar;

– Print feature: with this feature users can print maps and data on paper by selecting the print button placed in the menu bar;

– Edit—users can edit existing layers if they need to change or add something. This feature can be activated by clicking on the button Edit placed in the menu bar;

– Export—users can save visualized data as a picture file. This feature can be activated by clicking on the button Export placed in the menu bar;

– Links—users can open a predefined link for weather maps, weather forecasts etc. This feature can be activated by clicking on the button Links placed in the menu bar;

– Information panel—give information for objects and objects features on the map (Fig. 14).

2. Flood module [2] (Fig. 15):

– Working with layers: they can be switched on or off very easily from the layer icon which opens the list with available layers and is placed on the top right part of the screen or the base layers can be changed based on the user needs;

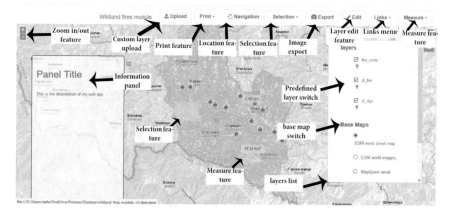

Fig. 14 Wildland fires module features

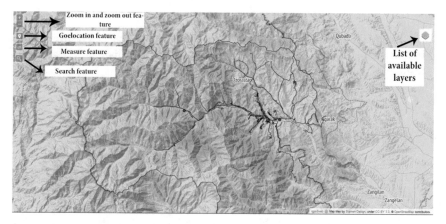

Fig. 15 Flood events module

- Zoom in and zoom out: can be used with mouse scroll or with the icons placed on the top left side of the screen;
- Geolocation: if users need to locate themselves they can use the icon placed on the top left side of the screen just under the zoom in and zoom out icons;
- Measures: if users need do measure distances on they can use the measure tool by clicking on the measure icon placed on the top left side of the screen just under the geolocation icon. After user activates this feature by clicking on the map, he/she can measure the distance between 2 or more points, for deactivating this feature the user needs to click on the measure icon;
- Searching: users can easily search different map objects by searching option or by clicking on the search icon placed on the top left side of the screen just under the measure icon and writing the needed text in the search field;

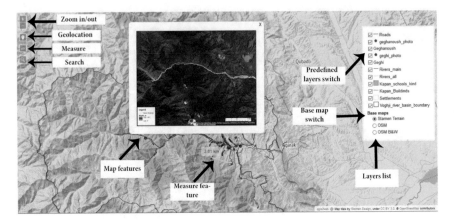

Fig. 16 Flood events module features

– Map features: when users need a different data features (example: orthophotos, information about cities, buildings and etc.) on the map they can explore this features by clicking on objects placed on the map (Fig. 16).

5 Tests and Expected Future Development

The wildfires module [1] and flood module [2] can be tested on field works within the frameworks of two international projects eOUTLAND and ALTER. The two modules can be unified into one decision support application with more ICT tools that will support work on the field [1] in different natural hazards. The application can deliver more detailed and complex information which could be used for field support to firefighter and volunteer groups in cases of wildland fires or flood events.

Our idea is to develop third module as part of the decision support application. This module will be designed to support activities in cases of flood events located in the area of Svilengrad municipality very close to the Bulgarian-Greek-Turkish border. The other future improvement of the application development is to visualize real time open source data from Copernicus satellite program of European Union. When we add all features and improvements the application will deliver more data and information supporting groups on field in cases of natural disasters.

6 Conclusion

This article presented an idea for desktop web based decision support application developed by usage of free and open source data and software solutions. Its potential users are volunteer groups and firefighting teams that need decision support tools in operational rooms [1]. This kind of ICT solutions are necessary for the growing needs in the context of wildland fires and flood events. Test areas for this tool and activities are Zlatograd, Madan, Nedelino municipal areas in Bulgaria and in Armenia are the Akhtala and Teghut areas of Lori Marz along the Shamlugh river, the Vorotan Cascade and its associated dams in the Syunik region, and the Kapan and Voghji river basin of Syunik region. We are using services, based on the latest GIS technologies adapted for building this kind of platforms for management of large data sets. This kind of tools also gives opportunity to firefighters and volunteer teams, in fire departments and operational rooms to react faster in cases of wildland fires and flood events. The main goal of the application is to deliver information about Points of Interest (POIs) which interested groups need to know like logistic centers for water supplies and firefighting tools, vulnerable buildings etc. With such tool the operational teams do not need to have in their pockets paper instructions and paper maps to fulfill their field work.

Acknowledgements This paper has been supported partially by the Bulgarian National Science Fund project number DN 12/5 called: Efficient Stochastic Methods and Algorithms for Large-Scale Problems and the DG ECHO project called: Alliance for disaster Risk Reduction in Armenia with acronym: ALTER and Grand Number: 783214.

References

1. Garca, V.G., Perotto-Baldivieso, H.L., Hallett, S.: A Prototype Design for a Web-GIS Disaster Support System: the Bolivia Amazon Case Study (2010)
2. Das, A.K., Prakash, P., Sandilya, C., Subhani, S.: Development of web-based application for generating and publishing groundwater quality maps using RS/GIS technology and P. Mapper in Sattenapalle, Mandal, Guntur District, Andhra Pradesh. In: ICT and Critical Infrastructure: proceedings of the 48th Annual Convention of Computer Society of India, vol. 2, pp. 679–686 (2014)
3. Singh, H., Gambhir, D.: An open source approach to build a web GIS application. Int. J. Comput. Sci. Technol. (IJCST) **3**, p110–p112. ISSN: 2229-4333 (2012)
4. https://www.qgis.org/
5. http://boundlessgeo.github.io/qgis-plugins-documentation/webappbuilder/
6. https://sdk.boundlessgeo.com/docs/
7. http://geoserver.org/about/
8. https://www.postgresql.org/
9. https://openlayers.org/

Data Optimizations on Kresna Fire (2017) as Inputs for WFA Simulations

Nina Dobrinkova

Abstract Bulgaria is facing wildfires with increased intensity and high reoccurrence probability through years in the last three decades. Official analyses of the state presented that the burned areas in Bulgaria are 11,000 ha as average per year in the period 2005–2015. The most affected zones are in south-west and south-east parts of the country, where the regions of Blagoevgrad, Stara Zagora, Haskovo and Burgas are located. The fire which we will present in our paper is situated in the area of Kresna Gorge in south-west Bulgaria, Blagoevgrad region. The fire was active between 24–29 August 2017 and burned 1600 ha. It has been selected in order to be simulated as a first time Bulgarian case implemented in the Wildfire Analyst (WFA) tool. WFA is computer based decision support system applicable for post fire or real time fire propagation analysis. This fire simulation data has been prepared in a way that future users can optimize their preliminary steps and directly apply the decision support methods in WFA tool.

1 Introduction

In Bulgaria Fire Behavior Fuel Modeling (FBFM) is not applicable as usual practice to forest maps and forest data on state, regional or local levels. This gives many opportunities to the fire analysts to elaborate the data in the best possible way. However that results in time consuming work which makes computer based tools hard to use in real time test cases. The work presented will give step by step data preparation

---

N. Dobrinkova (✉)
Institute of Information and Communication Technologies, Bulgarian Academy of Sciences, acad. Georgi Bonchev bl. 2, Sofia, Bulgaria
e-mail: ninabox2002@gmail.com; nido@math.bas.bg

© Springer Nature Switzerland AG 2020
S. Fidanova (ed.), *Recent Advances in Computational Optimization*,
Studies in Computational Intelligence 838,
https://doi.org/10.1007/978-3-030-22723-4_3

for simulation needs with Wildfire Analyst (WFA; [1]). We have chosen the Kresna fire as a case study, because it was one of the most destructive fires on Bulgarian territory. We used Anderson and Scott–Burgan [2, 3] as base models for the simulations. We applied the fuel load and descriptions given in each model explanation in order to adapt the data in the Bulgarian forestry maps polygons and sub-polygons in the affected area. The total elaborated data divided in FBFM models was for more than 2300 ha. and we used the GIS tolls to adapt the forestry maps to our needs. The fuel model descriptions which we had about the forest fire behavior included the following set of input data:

- Cumulative amount of combustible material (green and dry plants).
- Three classes according to the size of the combustible material (0–0.5, 0.5–2.5, 2.5–7.5 cm. in diameter).
- Surface—area—to—volume (SAV)—the accumulated flammable material per unit area.
- Accumulate energy released in the form of combustion heat according to the size of the fuel materials.
- The accumulated amount of combustible material in the fire area.
- Humidity of dead fuel accumulated from previous seasons per unit area.

This division was very important because we had to use pre-fire data from local people descriptions and some data had to be estimated according to our best judgment based on satellite images or orthophoto before the fire start.

The wildfire picks in Bulgaria are in the periods early spring, between March and April; or in the summer between July and October. 2017 was extreme year from point of view of wild land fires, because many places which do not really catch fires have been affected. Kresna gorge was one of them. It is important because of its geographical location along Struma Motorway. At this part the gorge highway is part of the EU corridor IV connecting Vidin (North-west Bulgaria) and Thessaloniki (North-west Greece) areas. The estimated losses after that fire for the Bulgarian country were about 15 million leva (approx. 7.5 million EUR), with 7 houses totally burned down, 1 destroyed car and 16,000 decars (1,600 ha), including large Natura 2000 protected zones. During the fire, the Trans-European corridor IV was closed and not operational, triggering high economic losses for transport companies and local population. In our paper we will present in different sections what type of data has been collected how it has been elaborated and than what kind of steps were followed in order to be done a successful WFA simulation. All of this steps if done preliminary in the forestry departments with the use of the nowadays ICT tools and WFA as fire propagation simulator can give local authorities opportunity for faster decision making in cases of wild land fires.

2 Data Preparation Steps Before Fire Spread Simulation to Be Done with WFA

Automated systems for detection of forest and field fires worldwide are divided into four main groups:

(a) Surveillance systems with cameras,
(b) Ground-based surveillance systems without cameras, but only with smoke and heat sensing sensors,
(c) Unmanned and man-operated airplanes or drones to transmit directly information on the presence of forest or field fire,
(d) Satellite monitoring.

Aviation tolls and satellite surveillance are over costly methods for day-to-day monitoring, and many countries around the world prefer the first two approaches. Another disadvantage for both satellite and aviation tools methods is the distortion of information in dense clouds or other weather extremes.

Terrestrial non-camera monitoring systems with data transfer sensors are a good and cost-effective option but do not transmit a picture at the command center. This makes them less preferable for operational needs. The ground systems that have cameras for real picture and data transmission are better option and the most used ones. Therefore, off-site terrestrial observation systems using CCD (Charged Coupled Device) sensing sensors in the visible and near-infrared spectra are the best and most cost-effective solution for implementing automatic surveillance and detection of forest fire with available technological solutions at the moment. In almost every country where there is a high risk of forest fires, at least one such system has been developed and proposed. The popular and common name of this spotting wildfire method is "tower". This is the most used tool on the Bulgarian territory for wildfire detection.

After a fire is detected by the tower solutions in the vulnerable area the coordinates or the ruff estimation of the source of the fire is sent in operational room or to a responsible authority's facility. In this case the option for the firefighters send on the field is to have decision support tool with options to simulate the probable fire spread. Such algorithm is presented in Fig. 1.

In our case study for post fire reconstruction this algorithm is fulfilled by Wildfire Analyst (WFA) tool. All received data about Kresna fire have been evaluated and processed. The input data for the tool was divided in three general types.

The first type was the meteorological data. In this data set we had to have information about: wind direction, wind speed, relative humidity, air temperature, calculated moisture, fine dead fuel moisture (1 h), live woody fuel moisture, Live herbaceous fuel moisture, cloud cover, date and time, Medium dead fuel moisture (10 h), Heavy dead fuel moisture (100 h). WFA gives opportunity to the user to use four options of weather data sets based on the listed parameters and needs—if it is real time fire propagation or past event reconstruction simulations:

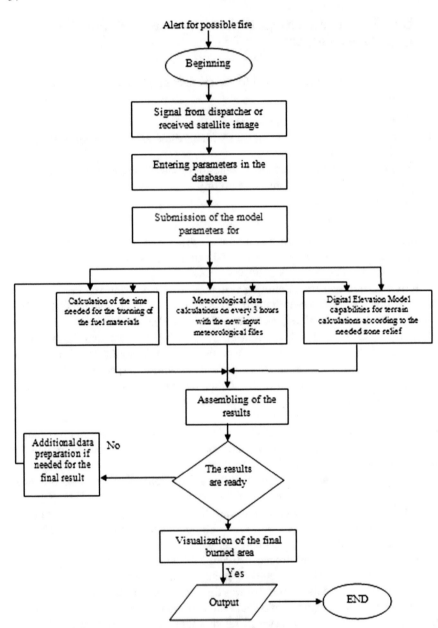

Fig. 1 Fire simulator algorithm processing the incoming signal

(a) Constant weather data—For the constant data option the weather data input is constant for the whole duration of the simulation. The constant weather data options show input variables, measurement units, minimum and maximum thresholds as well as the cloudiness status.

(b) Weather scenarios—The weather scenario functionality allows users to define the different weather inputs for each hour of the simulation duration. Different situations can be created according to the user knowledge or the characteristics of a particular area.

(c) Weather reconstruction data—In this option the user has to define a scenario based on collected historical weather data to reconstruct a wildfire simulation. Each scenario must have at least one simulation hour. Its working manner is similar to the one used in the weather scenarios, however in this case the user can set the weather conditions for more than 24 h.

(d) Predictions web-service—Another way of setting the weather data is to obtain it through a weather forecast service, which is an option to select from drop down menu in the WFA tool. It connects the user to the weather forecast server.

The second type of data was the Geographic Information Systems (GIS) layers. They are needed because this is the only way how we can put in easy and user-friendly way the exact location on the earth of our starting fire. In order to simulate the fire in Kresna gorge we had to create a special data set of layers in GIS format needed for the simulation purposes. We have created:

Vector data set for the Kresna Fire simulation purposes

1. Polygon layer representing Blagoevgrad region.
2. Polygon layer representing the municipalities within the research area.
3. Point layer representing the settlements within the research area.
4. Polygon layer representing the land within the research area.
5. Polygon layer of cadastral properties within the land areas wholly or partially within the research area.
6. Polygon layer presenting the scope of a survey area.
7. Polygon layer of the extent of the burnt area to 27.08.2017.
8. Polygon layer of the extent of the burnt area to 01.09.2017.
9. Point layer of temperature anomalies showing active fire from satellites MODIS and VIIRS.
10. Polygon layer representing Burned area index.
11. Polygon layer representing Normalized Burn Ratio.
12. Polygon layer representing water bodies—dams, ponds, etc.
13. Linear layer representing rivers.
14. Polygon layer representing the Earth's coverage within a surveyed region according to the classification CORINE LAND COVER.
15. Polygon layer representing the land cover within a surveyed area according to the classification of the Land Parcel Identification System (LPIS).
16. Polygon layer with the contours from the forest management plans of the State forestry department in Kresna and State forestry department Simitli.

17. Polygon layer Inscriptions from the forest management plans of the State forestry department in Kresna and State forestry department Simitli.
18. Polygon layer Forest departments and divisions on the territory of State forestry department in Kresna and State forestry department Simitli.
19. Polygon layer Signs from the forest management plans of the State forestry department in Kresna and State forestry department Simitli.
20. Linear layer representing railway lines within the research area.
21. Linear layer representing the roads within the research area.
22. Taxonomic description of the forest departments and sub—divisions in the State forestry department in Kresna and State forestry department Simitli.
23. Taxonomic description of the forest departments and sub—divisions in the State forestry department in Kresna and State forestry department Simitli.

Raster data set for the Kresna Fire simulation purposes

1. Aerophoto data set with 16 aerophoto images captured in 2013.
2. Digital Elevation Model (DEM) of the area.
3. Aspect on slope with cubicle size 25 m.
4. Digital model of relief with a cell side size of 25 m.
5. Hillsgade with a cell side size of 25 m.
6. Slope (degrees) with cell side size 25.

Satellite Images data set for the Kresna Fire simulation purposes

7. An image of Sentinel-2 shot on 22.08.2017 in 4 spectral channels (red, green, blue, infrared) was used for a reference.
8. An image of Sentinel-2 shot on 27.08.2017 was used for a reference.
9. An image of Sentinel-2 shot on 1.09.2017 was used for a reference.
10. An image of Sentinel-2 shot on 09.04.2018 was used for a reference.

The third type of data which was needed for fire simulation with WFA was the fuel types divided by applying the general rules of Anderson and Scott–Burgan models [2, 3]. In order to be described all fuel types as accurately as possible vector data sets have been created:

1. Polygon layer of the earth's surface in the area affected by the fire including its adjacent territory within a radius of 1 km.
2. Polygon layer of range of affected by the fire, including its adjacent territory within a radius of 1 km.
3. Polygonal layer presenting the scope of the territory survey. It covers the entire burnt area as well as its adjacent 1 km radius.
4. Polygonal layer land_fireCover created by combining the forest management plans of Kresna and Simitli forestry departments (layer lesso_Parcel is part from the main database) and a polygon layer presenting the mode of sustainable use on the territory according to the Land Parcel Identification System—kayer land_LPIS from the main Database.

Table 1 Fuel nomenclatures used in the attributive table where the fuel data is described

Name of the field	Description
SOURCE	Source of the classification system: 1. Forestry department plan, 2. Land Parcel Identification System
fuelType	The fuel types have been divided in two columns called Fire_Code—where Scott–Burgan types have been assigned and there is a second column called FIRE_Ander, which is the Anderson 1982 clssification. In the Anderson column have been put 0 as non burning areas like rocks, houses, roads, water etc.

When created the layer land_fireCover have been followed the rules:

(a) In the layer were included all sub-sections of the forest management projects which were wholly or partially located in the Burned Area of the fire.
(b) The territory that was not covered by the forest sub-sections is filled with information about the mode of sustainable use on the territory according to the Land Parcel Identification System.
(c) Contact areas between sub-divisions of forest management projects and the mode of sustainable use on the territory according to the Land Parcel Identification System have been edited. In this revision, the original appearance of the sub-sections was not changed.

Special table about fuel layer was created with the name fire_N_FuelType which was a nomenclature of the land cover types in a field FUEL_COVER_ver01 from layer land_fireCover. These fuel types are summarized in Table 1.

3 Wildfire Analyst (WFA) Tool and Its Application in Bulgaria

The Wildfire Analyst (WFA) tool is a software that can provide real-time analysis of wildfire behavior and simulates the spread of wildfires (Fig. 2). Simulations can be done fast in seconds I by using regular PC, Laptop or Tablet. It is software with capability to support real time decision making. WFA goal when designed was to address fast analysis for a different range of situations and users. WFA provides integration with ESRI's ArcGIS with no conversion or pre-processing required. This is added value from point of view of usability, which is allowing users to concentrate on interpreting simulation outputs, and making important decisions about how and where to deploy firefighting resources. The can be used at incident command center, in operations center, or directly on scene, providing outputs in less than a minute (Fig. 3). The software have as options to use predefined weather scenarios, or current and forecasted weather data obtained via web services. For wild land fire propagation, time is of the essence, and WFA architecture is done in a way to support initial attack situations, giving the Fire Chief and Incident Commander the critical intel-

Fig. 2 Simulation of a fire in the US with Wildfire analyst

ligence needed to support resource allocation decision making. This fire simulator tool provides a range of analytical outputs, available as GIS maps and charts for more accurate and timely decision making. WFA is a software component of Technosylva's incident management software suite. Wildfire Analyst provides critical information to support the resource dispatching and allocation.

Contemporary fire behavior software tools required a high degree of specialization, training and effort in the preparation and conversion of GIS data to use the software. Historically, this has been a limitation in using these tools for initial attack and real time applications. This fast performance facilitates use of the outputs in real time and allows for constant adjustment based on field observations and deployment decisions by the incident team.

WFA has as range different fire spread models like Rothermel (1972) [4], Kitral (1998) [5] used in South America. It has embedded other models such as Nelson equations to estimate fuel moistures from weather data (2000) [6], WindNinja (Forthofer et al. 2009) [7]—software that computes spatially varying wind fields for wildland fire applications in complex terrain. The Input data of the software is similar to other fire simulators such FARSITE or BehavePlus, because all of them use the same fire spread models and equations.

WFA includes real time processing performance for automatic rate of spread (ROS) adjustments based on observations used to create fire behavior databases, calculation of evacuation time zones (or 'firesheds'), and integration of simulation results for asset and economic impact analysis [8, 9].The tool is used operationally in

Fig. 3 Simulation's results visualized on laptop, PC or tablets

some regions in Spain, CONAF and private companies in Chile, Italy or US wildfire services.

The Kresna test case which has been selected for the calibrations and experiments with WFA in Bulgaria had as parameters the following information:

- The time frame of the active fire was from 24–29 August 2017.
- The terrain of the fire was hilly and low-mountain, situated at elevations from about 250 to 750 m a.s.l.
- The fire affected the eastern slopes of Kresna gorge above Struma river located between the villages Mechkul (to the north) and Vlahi (to the south).
- The fire started in grassland-shrub zone and transferred to plantations from Pinus nigra, where it quickly spread assisted by the dry conditions.
- The total affected area was about 2260 ha, of which 65% were forest territories.
- The fire burnt mostly plantations from Austrian pine (870 ha), Scots pine (Pinus sylvestris), Robinia pseudoacacia (21,1 ha), natural forests dominated by Quercus pubescens (200 ha) and Quercus petraea (66 ha) and smaller patches of other species. Pine plantations were very affected by fire with rate of mortality higher than 80% due to a high burn severity.

- On steeper slopes above Struma River the severity of burning was lower, where some deciduous trees and single specimen of Juniperus excelsa survived.
- The affected zones were mostly protected ones under the NATURA 2000 areas.

The initialization of WFA has been done with predefined layers in GIS format using the logic presented in the three types of data sets preliminary prepared prior the simulations start. All data was collected from different sources. The fire ignition point was selected after careful evaluation of all materials and conversation with the locals who took part in the fire suppression measures.

For a reference have been used satellite data from VIIRS. This was a way to compare the official reports from the responsible authorities and the real fire spread captured by the satellite images (see Fig. 4).

Data samples about fuels and first runs with WFA of the Kresna fire have been presented on Figs. 5 and 6. The data used in the initial simulations was with buffer area around the real fire perimeter of 1 km. The next set of data preparation includes buffer zone of 10 km. This two different sizes of buffer zones give an opportunity for simulations that can be done in a way using different propagation scenarios. For analyses purposes this is very useful approach and can give to the local authorities post fire propagation information of its reasons to evolve in time.

WFA tool among its tools have capacity for the so called adjustment option. This is a special ability of the tool during the fire simulations in real-time or when past event reconstruction is in place adjustment of the fire spread to be inserted during the simulation calculations. This option is possible by identification of adjustment points (also often mentioned as control points) of the fire based on observations from the field. However only people with very good understanding of wildfire spread can make accurate adjustment predictions. Thus adjustment capability main usage is feasible reconstruction of past or historical fire events is in place. WFA automatically allows the adjustment of the ROS of existing fuels and is given by the following formula:

$$ROS = ADJUSTMENT \ X \ ROS \ (humidity, fuel, wind, slope)$$

The adjustment in the formula is represented by a certain value and the ROS is the rate of spread of the fire which is influenced by several factors during its calculation such as the humidity, fuel, wind, and slope of the terrain.

That option of the tool was evaluated and tested with the 1 km buffer set of data. The results and the new data sets of 10 km buffer data could be used for future table top exercises in the south-west Bulgaria in order local authorities, firefighters and volunteer groups to test their decision making procedures with the WFA simulator. WFA as tool is very powerful. Its capabilities for historical reconstruction of past events plus application the fire spread models implemented as libraries or modes for simulations in real time are very useful.

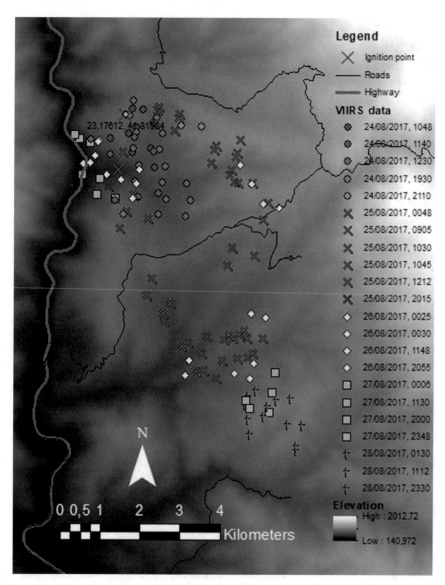

Fig. 4 Kresna fire progression through VIIRS data

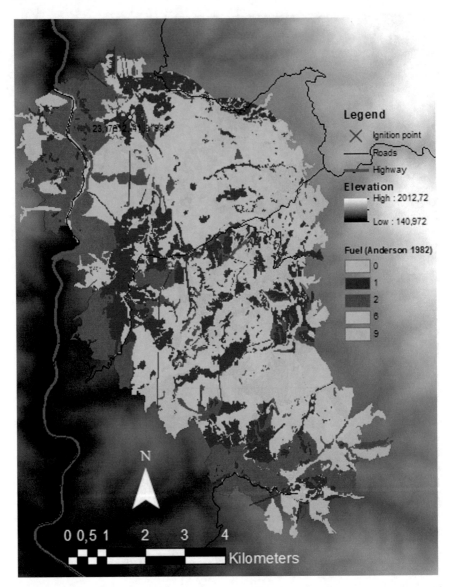

Fig. 5 Kresna fire fuel map

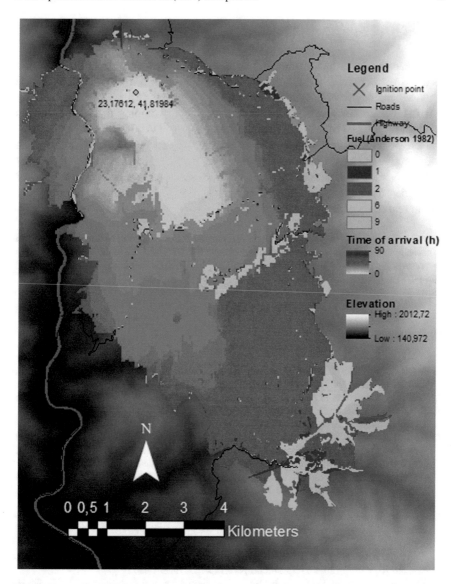

Fig. 6 Kresna fire WFA output with 1 km buffer zone around the burned area

4 Conclusion

The presented work has as aim to provide to the interested readers how to build its data in a way suitable for simulations with Wildfire Analyst (WFA) tool. As we could see major part of the work flow is the data preparation and its design in a way that all layers follow common logic for the specific area of interest. This approach is good for small areas of interest where fast analyses need to be done. For bigger areas and use of WFA in operational capacity all forestry data and data outside the urban areas need to be elaborated following the presented processing steps. The adjustment factor capability and the accurate meteorological data can give good option for local authorities, firefighters and volunteer groups to use WFA as real time decision support tool in real cases of fire propagation.

Acknowledgements This work has been partially supported by Bulgarian National Science Fund project number DN 12/5 called: Efficient Stochastic Methods and Algorithms for Large-Scale Problems and the Interreg Balkan-Mediterranean project "Drought and fire observatory and early warning system—DISARM".

References

1. Cardil, A., Molina, D.M.: Factors causing victims of wildland fires in Spain (1980–2010). Hum. Ecol. Risk Assess. Int. J. **21**, 67–80 (2015). https://doi.org/10.1080/10807039.2013.871995
2. Anderson, H.E.: Aids to determining fuel models for estimating fire behavior. USDA Forest Service, Intermountain Forest and Range Experiment Station, Research Report INT-122 (1982). http://www.fs.fed.us/rm/pubsint/intgtr122.html
3. Scott, J.H., Burgan, R.E.: Standard fire behavior fuel models: a comprehensive set for use with Rothermel's surface fire spread model. General Technical report RMRS-GTR-153. Fort Collins, CO: U.S. Department of Agriculture, Forest Service, Rocky Mountain Research Station. 72 p
4. Rothermel, R.: A mathematical model for predicting fire spread in wildland fuels. USDA Forest Service, Intermountain Forest and Range Experiment Station, (Ogden, UT). Research Paper INT-115 (1972)
5. Pedernera, P., Julio, G.: Improving the economic efficiency of combatting forest fires in chile: the KITRAL system. USDA Forest Service General Technical report PSWGTR-173, pp. 149–155 (1999)
6. Nelson Jr., R.M.: Prediction of diurnal change in 10-h fuel stick moisture content. Can. J. For. Res. **30**, 1071–1087 (2000). https://doi.org/10.1139/CJFR-30-7-1071
7. Forthofer, J.; Shannon, K.; Butler, B.: Simulating diurnally driven slope winds with WindNinja (link is external). In: Proceedings of 8th Symposium on Fire and Forest Meteorological Society, 13–15 October 2009, Kalispell, MT (2,037 KB; 13 pp.) (2009)
8. Ramirez, J., Monedero, S., Buckley, D.: New approaches in fire simulations analysis with Wildfire Analyst. In: The 5th International Wildland Fire Conference. Sun City, South Africa (2011)
9. Monedero, S., Ramirez, J., Molina-Terrén, D., Cardil, A.: Simulating wildfires backwards in time from the final fire perimeter in point-functional fire models. Environ. Model. Softw. **92**, 163–168 (2017)

Solving Sorting of Rolling Stock Problems Utilizing Pseudochain Structures in Graphs

Jens Dörpinghaus and Rainer Schrader

Abstract Rearranging cars of an incoming train in a hump yard is a widely discussed topic. Sorting of Rolling Stock Problems can be described in several scenarios and with several constrains. We focus on the train marshalling problem where the incoming cars of a train are distributed to a certain number of sorting tracks. When pulled out again to build the outgoing train, cars sharing the same destination should appear consecutively. The goal is to minimize the number of sorting tracks. We suggest a graph-theoretic approach for this $\mathcal{NP}$-complete problem. The idea is to partition an associated directed graph into what we call pseudochains of minimum length. We describe a greedy-type heuristic to solve the partitioning problem which, on random instances, performs better than the known heuristics for the train marshalling problem. In addition we discuss the TMP with b-bounded sorting tracks.

1 Introduction

A hump yard usually consists of a hump and a set of classification or sorting tracks and one or more roll-in and pull-out tracks [1]. In hump yards freight cars are arranged or rearranged into a specific sequence of cars. The outgoing trains will deliver goods to new destinations. A practical introduction to hump yards with examples can be found in the work of Hiller [2].

This can be very complex. For example the hump yard in in Zürich-Limmattal (CH) consists of 18 roll-in tracks, 64 sorting tracks with a length of 650–850 m and 16 roll-out tracks, see [2, 3].

J. Dörpinghaus (✉)
Fraunhofer Institute for Algorithms and Scientific Computing,
Schloss Birlinghoven, Sankt Augustin, Germany
e-mail: jens.doerpinghaus@scai.fraunhofer.de

R. Schrader
Department for Mathematics and Computer Science,
University of Cologne, Cologne, Germany
e-mail: schrader@zpr.uni-koeln.de

© Springer Nature Switzerland AG 2020
S. Fidanova (ed.), *Recent Advances in Computational Optimization*,
Studies in Computational Intelligence 838,
https://doi.org/10.1007/978-3-030-22723-4_4

Every incoming car arriving at the hump yard will be assigned to a sorting track. At the end of this process all cars of every sorting track will be placed as a block on the roll-out track. For an optimization approach the number and length of sorting tracks, the number of roll-in and pull-out operations can be minimized.

Hansmann provided a general class of *Sorting of rolling Stock Problems* (SRSP) in [4]. We will focus on the *Train Marshalling Problem* (TMP): using a minimum number of tracks, rearrange the cars in a hump yard in such a way that cars sharing the same destinations appear consecutively in the rearranged train.

During the process only two movements are allowed: the sorting of cars to the tracks and one pull-out movement for every track. The tracks are not limited in length, so we can think of the tracks as stacks. We only allow one roll-in operation per car and one pull-out operation per track. No further shunting is allowed.

Apparently, TMP was first introduced by Zhu and Zhu [5] in 1983 who considered it under additional constraints and gave first results and polynomial algorithms. In 2000, Dahlhaus et al. [6] proved that TMP is $\mathcal{NP}$-complete and introduced new bounds. Brueggeman et al. show in [7] that the problem is fixed parameter tractable. In another work by Dahlhaus et al. [8] they described similar problems. More bounds and algorithms can be found in the work of Beygang [9] and Beygang et al. [1]. They introduced a graph-theoretic approach by considering the interval graph of a given instance. The problem also occurs in the work of Hansmann [4]. Other approaches can be found in the work of Rinaldi and Rizzi [10] who focused on dynamic programming and Haahr and Lusby [11].

First of all we will give a short formal problem description and all relevant definitions. After introducing pseudochains and discussion splittable destinations we will derive a novel greedy heuristic to solve the TMP. We will evaluate the results on some random instances and finish with a conclusion. This work goes back the results of [12]. After that we will transfer these results to a special variant of the TMP with limited sorting tracks.

2   Problem Description

With every *car i* in the hump yard we associate a natural number $\sigma_i \in \mathbb{N}^+$ representing the destination of the car. A *train* σ of length n then is a sequence

$$\sigma = (\sigma_1, \ldots, \sigma_n)$$

of cars with $\sigma_i \in \{1, \ldots, d\}$ for $i \in \{1, \ldots, n\}$.

Example 2.1 Let $\sigma = (1, 2, 1, 3, 2)$. There are three destinations, where the first and third car and, resp., the second and the last have the same destination.   □

We want to rearrange the cars in a departing train such that all cars are sorted in blocks according to their destination. As mentioned before, only two shunting

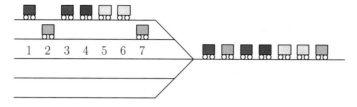

Fig. 1 Illustration for Example 2.2

operations are permitted: the roll-in movement of a car to one of the sorting tracks and the pull-out of all cars on a sorting track. The goal is to minimize the number of sorting tracks, denoted by $K(\sigma)$. Since only one shunting operation per sorting track is allowed the minimization of shunting operations is equivalent to the minimization of sorting tracks.

For a given sequence σ let S_k be the elements of σ with destination k. Then we may describe the incoming sequence by a partition $S = \{S_1, \ldots, S_d\}$ of $\{1, 2, \ldots, n\}$. Dahlhaus et al. [6] have shown that TMP may be rephrased as follows: find the smallest number $K(S)$ and a permutation π of $1, \ldots, d$ such that the sequence of numbers

$$\underbrace{1, 2, \ldots, n, 1, 2, \ldots, n, 1, 2, \ldots, n}_{K(S)\text{-times}}$$

contains the elements of $S_{\pi(1)}$ as a subsequence followed by all elements of $S_{\pi(2)}$ and so on.

Example 2.2 Let $n = 7$, $d = 4$ und $S = \{S_1, S_2, S_3, S_4\}$ with $S_1 = \{1, 3\}$, $S_2 = \{2, 7\}$, $S_3 = \{4\}$ and $S_4 = \{5, 6\}$. Then $K(S) = 2$ and $\pi(1, 3, 4, 2)$, see Fig. 1 for an illustration:

$$\underbrace{1\ 2\ 3}_{S_1}\ \underbrace{4}_{S_3}\ \underbrace{5\ 6}_{S_4}\ 7\ 1\ \underbrace{2\ 3\ 4\ 5\ 6\ 7}_{S_2}$$

We now define some necessary preliminaries following the work of Beygang in [9].

3 Preliminaries

Let $\mathbb{S}^n$ be the set of all problem instances of TMP with n cars. For $S \in \mathbb{S}^n$ let $d = d(S)$ be the number of destinations in this instance. For an instance $S \in \mathbb{S}^n$, a *track assignment* is function $tr : \{1, \ldots, n\} \to \mathbb{N}$ which assigns a track to every car. A track assignment is feasible if it gives a feasible solution for TMP.

For a given sequence σ and a destination k let $first(k)$ denote the position of the first occurrence of k and $last(k)$ be its last occurrence. Let $I_k = [first(k), last(k)]$

be the associated interval. Then the intervals induce a partial order on the set of destinations via $i < j$ if $last(i) < first(j)$. We consider the associated comparability graph and its complement, the interval graph.

Definition 1 (Comparability Graph associated with an Input Instance) For a given instance S of TMP, the associated comparability graph graph is given by $D(S) = (V, A)$ such that $V = (I_1, \ldots, I_d)$ and $(I_k, I_j) \in A$ if $k < j$. □

Definition 2 (Interval Graph associated with an Input Instance) For a given instance S of TMP, the associated interval graph is given by $G(S) = (V, E)$ such that $V = (I_1, \ldots, I_d)$ and $(I_k, I_j) \in E$ if $I_k \cap I_j \neq \emptyset$. □

Beygang already introduced some bounds and two important heuristics for the TMP. The deterministic SPLIT-Algorithm was introduced in [9] and computes a feasible solution by splitting destinations whenever possible in $O(n)$. The GREEDY-Algorithm was also introduced in [9]. It finds a feasible solution by partitioning the interval graph $G(S)$ into a minimum number $\chi(G_S)$ of stable sets, each assigned to one track. Recall that this is equivalent to partitioning $D(S)$ into a minimum number of chains. We will generalize this approach to partition $D(S)$ into pseudochains.

4 Pseudochains

Let $D = (V, A \cup B)$ be a directed graph with a set B of blue arcs, $A \cap B = \emptyset$. We allow that $B = \emptyset$ or $A = \emptyset$. Recall that a chain in a transitively oriented graph is a subset $v_1, \ldots, v_k$ of vertices such that $(v_i, v_j) \in A$ for all $1 \leq i < j \leq k$.

Definition 3 (Pseudochain) Let $D = (V, A \cup B)$ as above such that the subgraph D_A induced by the arcs in A is transitively orientable. $C \subseteq V$ is a *pseudochain* of length $\ell(C) = k \geq 1$ if C can be written as

$$C = C_1, b_2, C_2, b_3, C_3, \ldots, b_k, C_k$$

where the C_i's are mutually disjoint chains in D_A with last element a_i and first Element c_i and $(a_{i-1}, b_i), (b_i, c_i) \in B$ for $2 \leq i \leq k$. □

Figure 2 illustrates a pseudochain of length three.
Now we can define the minimization problem as follows.

Definition 4 (minPC) Given a directed graph $D = (V, A \cup B)$ with a set B of blue edges, $A \cap B = \emptyset$ such that D_A is transitively orientable. Partition V into pseudochains $P = C_1, \ldots, C_k$ such that the total length $\ell(P) = \sum_{i=1}^{k} \ell(C_i)$ of the partition is minimal. □

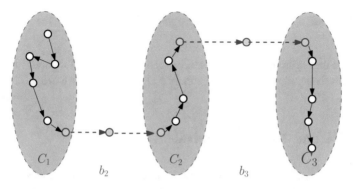

Fig. 2 A pseudostable chain of length 3. Transitive arcs are omitted, dashed arcs correspond to blue arcs

5 Splittable Destinations

Observe that we can always produce a feasible track assignment by opening a track for each destination. But we may be able to do better by distributing the cars of one destination to two tracks. For this, we consider three destinations (a, b, c) with $I_a \cap I_b \neq \emptyset$ and $I_b \cap I_c \neq \emptyset$. Thus we need at least two tracks for $S(a) \cup S(b)$ and two tracks for $S(b) \cup S(c)$. We call the destination b *splittable* with predecessor a and successor c if there is a feasible track assignment which assigns b to two different tracks. For short, we say that a triple (a, b, c) with the properties above is splittable.

In order to be feasible, some cars of $S(b)$ must then form a block at the end of one track containing the cars of $S(a)$ and a block at the beginning of some other track containig the cars of $S(c)$. It is easy to show the following Theorem:

Theorem 1 *(Splittable Destinations) Given an instance $S \in \mathbb{S}^n$ of TMP. Let (a, b, c) be a triple of destinations with $I_a \cap I_b \neq \emptyset$ and $I_b \cap I_c \neq \emptyset$. Then the triple (a, b, c) is splittable if and only if there is no car of destination b between $first(c)$ and $last(a)$.*   □

Proof Given a feasible track assignment and a destination b which is assigned to two tracks. We may assume that both tracks contain cars of other destinations. Let a be the destination preceding cars of b on the first track and c the destination following the cars on the other track. Then it is easy to see, that in the incoming train either $last(a) < first(c)$ or no car of destination b occurred between $first(c)$ and $last(a)$.

Conversely, let (a, b, c) a triple of destinations (a, b, c) as above. We start by assigning $S(a)$ to an open track, track 1 say, and cars of $S(c)$ to track 2. Cars of destination b will also be assigned to track 2 if they occur before $last(a)$, and to track one track 1 otherwise. All other cars are assigned to a destination-specific track. By assumption, either $last(a) < first(c)$, and the assignment is feasible. In the other case, after the first car of $S(c)$ is assigned to track 2, all remaining cars of $S(b)$ are assigned to the end of track 1. So in both cases the assignment is feasible.   □

Observe that splittable triples cannot be read off from the comparability graph D itself. So in the next section we will enhance D to capture this extra information.

6   minPC and TMP are Equivalent

Let D be the comparability graph of the intervals given by an instance $S \in \mathbb{S}^n$ and (a, b, c), a splittable triple. Observe that by definition I_b overlaps both I_a and I_c. So $(a, b), (b, c) \notin A$. Let $B = \{(a, b), (b, c) \mid (a, b, c)$ is a splittable triple$\}$ and $D^* = D^*(S) = (V, A \cup B)$ be the *extended comparability graph* of S.

Example 6.1 Given an instance $S \in \mathbb{S}^{50}$ with 16 destinations and

$$\sigma = (1, 1, 2, 1, 2, 3, 3, 3, 4, 2, 2, 1, 5, 3, 3, 4, 2, 1, 1, 6,$$
$$6, 2, 5, 7, 8, 1, 9, 10, 8, 11, 12, 13, 2, 5, 8, 10, 14, 14, 15, 16,$$
$$16, 12, 7, 4, 10, 5, 7, 8, 13, 11)$$

A partition P of the extended comparability graph $D^*(S)$ in pseudochains is given by

- $C_1 = \{14, 15, 16\}$ with $\ell(C_4) = 1$.
- $C_2 = \{1, 9, 10\}$ with $\ell(C_2) = 1$.
- $C_3 = \{2, 5, 7\}$ with $C_1 = \{2\}, b_2 = 5, C_2 = \{7\}$ and $\ell(P_1) = 2$.
- $C_4 = \{3, 4, 6, 8, 11, 12, 13\}$ with $C_1 = \{3, 13\}, b_2 = 11, C_2 = \{12\}, b_3 = 4, C_3 = \{6, 8\}$ and $\ell(P_3) = 3$.

See Fig. 3. The weight is $\ell(P) = 7$.   □

Lemma 6.2 *Let $S \in \mathbb{S}^n$ and P, a pseudochain partition of the extended comparability graph $D^*(S)$. Then P induces a feasible track assignment using $\ell(P)$ tracks.*   □

Proof It suffices to show that we can assign a pseudochain $C = C_1, b_2, C_2, b_3, C_3, \ldots, b_k, C_k$ of length k to k tracks. Let chain C_i begin with c_i and end with a_i. Let $B'_i = \{c \in b_i : c < last(c_{i-1})\}$ and $B\prime_i = \{c \in b_i : c > last(c_{i-1})\}$. We claim that, for $1 \leq i \leq k - 1$, we can schedule the pseudochain such that track i contains B'_{i-1} followed by C_i again followed by $B\prime_i$. Suppose this is true for some $1 \leq j < k$. Since, by definition, the triple (a_j, b_{j+1}, c_{j+1}) is splittable, we may fill track j with $B\prime_{j+1}$, open track $j + 1$ with B'_j and fill it with C_{j+1}.   □

Lemma 6.3 *Let $S \in \mathbb{S}^n$ and tr, a feasible track assignment using k trains. Then tr induces a pseudochain partition P of the extended comparability graph with $\ell(P) = k$.*   □

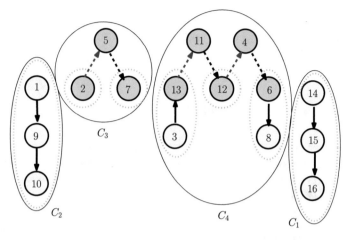

Fig. 3 A pseudochain partition found in Example 6.1

Proof Since tr is feasible, the cars of a destination d are assigned to at most two tracks and form a consecutive subsequence on their tracks. If they are assigned to two tracks they must be placed at the end of one track and at the beginning of some other track. Define a directed graph H on the set of destinations. Two destinations i, j are linked by an edge (i, j) if cars of i are placed immediately before cars of j on the same track. Then each connected component of H induces a pseudochain C of $D(S)$. Since $\ell(C)$ corresponds to the number of tracks used by the component, the claim follows. $\qquad\square$

Example 6.4 Consider the instance of Example 6.1. The pseudochain partition P induces the following track assignment:

$$
\begin{aligned}
\text{Track 1} &: 2_3, 2_{33}\ 5_{34}, 5_{51} \\
\text{Track 2} &: 5_1, 5_{23}\ 7_{24}, 7_{47} \\
\text{Track 3} &: 1_1, 1_{26}\ 9_{27}\ 10_{28}, 10_{45} \\
\text{Track 4} &: 3_6, 3_{15}\ 13_{32}, 13_{49}\ 11_{50}, 11_{51} \\
\text{Track 5} &: 11_1, 11_{30}\ 12_{31}, 12_{42}\ 4_{44}, 4_{51} \\
\text{Track 6} &: 4_9, 4_{16}\ 6_{20}, 6_{21}\ 8_{25}, 8_{48} \\
\text{Track 7} &: 14_{37}, 14_{38}\ 15_{39}\ 16_{40}, 16_{41}
\end{aligned}
$$

Here, the lower indices represent the position of the car in the input sequence. It is a feasible track assignment using $\ell(P) = 7$ tracks. $\qquad\square$

Theorem 2 *Let $S \in \mathbb{S}^n$ and $D^*(S)$, the extended comparability graph. Then minPC on $D^*(S)$ is equivalent to TMP.* $\qquad\square$

Proof Follows from Lemma 6.2 and 6.3. $\qquad\square$

Thus for every optimal solution of an instance $S \in \mathbb{S}^n$ of the TMP using $K(S)$ tracks, there exists a corresponding partition of the extended comparability graph $D(S)$ into pseudochains.

7 A New Greedy-Approach: Greedy-PC

This greedy approach is based on the above observations on pseudochain partitions. Let $S \in \mathbb{S}^n$ be an instance of TMP and $T = (t_1, \ldots, t_{t_S})$ be the list of all splittable destination in S sorted increasingly by their left boundary. Our approach will return a partition of $D^*(S)$ into pseudochains.

Algorithm 1 GREEDY-PC

Require: Extended comparability graph $D^*(S)$ with its maximal stable set S in $D(S)$ and a list
$\quad T = (t_1, \ldots, t_{t_S})$ of splittable destinations in S.
Ensure: Partition of $D^*(S)$ in pseudochains
1: $visited = \emptyset$
2: $count = 0$
3: **while** $|T| > 0$ **do**
4: $\quad count + +$
5: $\quad P.add(\text{pseudochain } P_{count})$
6: $\quad$ **for** every $(a, b, c) = t_i \in T$ **do**
7: $\quad\quad$ **if** $P_{count}.add(a, b, c) = true$ **then**
8: $\quad\quad\quad visited.add\ a, b, c$
9: $\quad\quad$ **end if**
10: $\quad$ **end for**
11: $\quad$ **for** every $v \in visited$ **do**
12: $\quad\quad$ delete every t_i containing v from T
13: $\quad$ **end for**
14: **end while**
15: **for** every node $v \in V(G)$ **do**
16: $\quad$ **if** $v \notin visited$ **then**
17: $\quad\quad$ **for** $i = 1, \ldots, count$ **do**
18: $\quad\quad\quad$ **if** $P_i.add(v) = true$ **then**
19: $\quad\quad\quad\quad visited.add\ v$
20: $\quad\quad\quad\quad exit$
21: $\quad\quad\quad$ **end if**
22: $\quad\quad$ **end for**
23: $\quad\quad count + +$
24: $\quad\quad P.add(\text{pseudochain } P_{count})$
25: $\quad\quad P_{count}.add(v)$
26: $\quad$ **end if**
27: **end for**
28: **return** P

Given a splittable destination $(a, b, c) \in T$, the function add applied to a pseudochain P returns `true` if the triple can be added to P and `false` otherwise. We follow the idea to have the best solution within this chain P and try to add every possible splittable triple in T to a pseudochain. We will redo this as long as nodes remain in S.

The worst-case runtime of this heuristic is $f(n) = (\frac{1}{2}n^3 + n^2) = O(n^3)$. See Algorithm 1 for an implementation in pseudocode. This problem was mentioned in the work of [4], but

8 Experimental Results

We used Python 3.4 with NetworkX for creating random instances and implement the greedy heuristic as well as the Linear Programming relaxation introduced by Beygang [9]. Four 2.4 GHz processors and 8 GB RAM were available running Linux Kernel 3.10. We used GLPK (GNU Linear Programming Kit) 4.52 to solve the linear program. To get comparable results, we followed [9] to create random instances. This function takes the number n of cars and computes uniform and independent problem instances.

The greedy heuristic introduced by Beygang et al. [1, 9] leads to the upper bound denoted by *Coloring*. It is equivalent to a graph coloring approach for the interval graph $G(S)$. The runtime is $O(n^2)$. Algorithm 1 has also polynomial runtime in $O(n^3)$.

We approximate the optimal solution according to the bounds u_{lp} and l_{lp}, the upper and lower bound given by the solution of the linear program introduced by [9]. It was observed in [13] that the lower bound very often coincides with the value of the optimal solution.

Figures 4 and 5 summarize the output of the heuristics on 50 random instances with a fixed number of cars. For a small number of cars the distance between *Coloring* and u_{greedy} is small, but notable, see Fig. 4. We notice the greedy approach can lead to solutions using more tracks than the *Coloring* approach. The situation changes significantly for instances with more cars. Figure 5 shows that Greedy-PC performs better than *Coloring* on instances with 300 cars.

9 TMP With b-Bounded Sorting Tracks

We will now transfer the results of the previous sections to a special version of the TMP where the length of sorting tracks is bounded to b cars. Even recent research like [14] did not consider this case since it is more complex than the unbounded version. We will show that we can use weighted pseudochains to find a graph-theoretic equivalent to this problem.

Let $D = (V, A \cup B)$ such that the subgraph D_A induced by the arcs in A is transitively orientable. Recall that every Pseudochain has the form

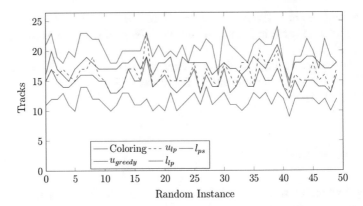

Fig. 4 Results for random instances with $n = 100$ cars. *Colouring* is a Greedy-approach introduced by Beygang, u_{lp} and l_{lp} are upper and lower bounds of the integer linear program approach, see [9]. The lower bound l_{ps} was introduced in [13]. u_{greedy} shows the results of our novel Algorithm 1

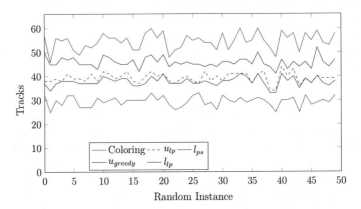

Fig. 5 Results for random instances with $n = 300$ *Colouring* is a Greedy-approach introduced by Beygang, u_{lp} and l_{lp} are upper and lower bounds of the integer linear program approach, see [9]. The lower bound l_{ps} was introduced in [13]. u_{greedy} shows the results of our novel Algorithm 1

$$C = C_1, b_2, C_2, b_3, C_3, \ldots, b_k, C_k,$$

with C_i's mutually disjoint chains in D_A with last element a_i and first Element c_i and $(a_{i-1}, b_i), (b_i, c_i) \in B$ for $2 \leq i \leq k$.

Now we can define the weight of every disjoint chain $C_i = \{c_1, \ldots, c_n\}$ with it's precursor b_i and successor b_{i+1} as

$$w(C_i) = \sum_{i=1}^{n} w(c_i) + w_l(b_{i-1}) + w_r(b_i)$$

According to the definition of the TMP we set $w(c)$ as the numbers of cars with destination c. If three destinations (a, b, c) are splittable we set $w_l(b)$ as the numbers

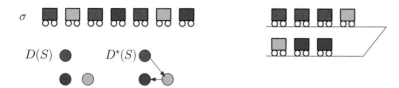

Fig. 6 An incoming train $\sigma = \{1, 2, 1, 3, 1, 2, 3\}$. At the bottom illustrations of the graphs $D(S)$ and the extended comparability graph $D^*(S)$. On the right the example track assignment given by the pseudochain partition $C = \{a, b, c\}$ of $D^*(S)$

of cars of destination b assigned to a sorting track before this sorting track is closed and $w_r(b)$ for the number of cars after this event.

Given a feasible track assignment solving an instance $S \in \mathbb{S}^n$. With this definition we can count the number of cars on each sorting track. Thus we can now define an additional bound to solve the b-bounded version of the TMP.

Let $S \in \mathbb{S}^n$ and P, a pseudochain partition of the extended comparability graph $D^*(S)$ with

$$m(P) = \max_{C_i \in P} w(C_i) \leq b$$

Algorithm 2 CLEAN-B-BOUNDED

Require: A solution P of a TMP instance $S \in \mathbb{S}^n$ and $b \in \mathbb{N}$. we can see that if $m(P) \leq b$ this is also a valid solution of b-bounded TMP.
Ensure: Solution P of a b-bounded TMP instance $S \in \mathbb{S}^n$
1: compute $m = m(P)$
2: **if** $m \leq b$ **then**
3:    return P
4: **end if**
5: $D = \emptyset$
6: **for** every $C_i \in P$ with $w(C_i) > b$ **do**
7:    Remove c_i in C_i for $i = \{1, \ldots, n\}$ until $w(C_i) \leq b$
8:    add all removed c_i to D
9: **end for**
10: add every $c_i \in D$ to a new chain in P
11: **return** P

Then P induces a feasible track assignment using $\ell(P)$ tracks with at most b cars per track.

Example 9.1 See the example train $\sigma = \{1, 2, 1, 3, 1, 2, 3\}$ in Fig. 6. Here $m(P) = 4$, thus this is a valid solution for every b-bounded version of the TMP with $b \geq 4$. A feasible track assignment would be

$$\text{Track } 1 : 1_1; 1_3; 1_5; 2_6$$
$$\text{Track } 2 : 2_2; 3_4; 3_7$$

If we set $b = 3$ this is no longer a feasible track assignment. Thus a feasible solution of 3-bounded TMP would need three sorting tracks with the track assignment:

$$\text{Track 1} \; : \; 1_1; \, 1_3; \, 1_5;$$
$$\text{Track 2} \; : \; 2_2; \, 2_6$$
$$\text{Track 3} \; : \; 3_4; \, 3_7$$

□

To solve any instance of b-bounded TMP we can use two strategies. Given a solution P of a TMP instance $S \in \mathbb{S}^n$ we can see that if $m(P) \le b$ this is also a valid solution of b-bounded TMP. If there is a $C_i \in P$ with $w(C_i) > b$ we can remove all destinations $c \in C_i$ till $w(C_i) \le b$. This procedure can be applied for all $C_i \in P$ with

---

Algorithm 3 GREEDY-PC-B-BOUNDED

Require: Extended comparability graph $D^*(S)$ with its maximal stable set S in $D(S)$ and a list
    $T = (t_1, \ldots, t_{t_S})$ of splittable destinations in S and $b \in \mathbb{N}$.
Ensure: Partition P of $D^*(S)$ in pseudochains with $m(P) \le b$
1: $visited = \emptyset$
2: $count = 0$
3: **while** $|T| > 0$ **do**
4:     $count + +$
5:     $P.add(\text{pseudochain } P_{count})$
6:     **for** every $(a, b, c) = t_i \in T$ **do**
7:         **if** $P_{count}.add(a, b, c) = true$ and $P_{count}.weight \le b$ **then**
8:             $visited.add\ a, b, c$
9:         **end if**
10:     **end for**
11:     **for** every $v \in visited$ **do**
12:         delete every t_i containing v from T
13:     **end for**
14: **end while**
15: **for** every node $v \in V(G)$ **do**
16:     **if** $v \notin visited$ **then**
17:         **for** $i = 1, \ldots, count$ **do**
18:             **if** $P_i.add(v) = true$ and $P_i.weight \le b$ **then**
19:                 $visited.add\ v$
20:                 $exit$
21:             **end if**
22:         **end for**
23:         **if** $w(v) \le b$ **then**
24:             $count + +$
25:             $P.add(\text{pseudochain } P_{count})$
26:             $P_{count}.add(v)$
27:         **else**
28:             **for** every part $v_i \in v$ with $w(v_i) \le b$ **do**
29:                 $count + +$
30:                 $P.add(\text{pseudochain } P_{count})$
31:                 $P_{count}.add(v_i)$
32:             **end for**
33:         **end if**
34:     **end if**
35: **end for**
36: **return** P

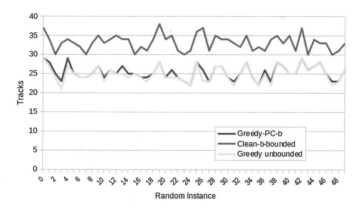

Fig. 7 Results for random instances with $n = 150$ cars solving 5-bounded TMP. We use a u_{greedy} introduced as novel heuristic to solve unbounded TMP to compare the results of b-bounded to unbounded TMP. GREEDY-PC-B-BOUNDED is an adjusted version of this heuristic, CLEAN-B-BOUNDED tries to find a feasible solution by fixing the unbounded solution

$w(C_i) > b$. Then we can find a minimum solution for the removed destinations. See Algorithm 2 for pseudocode.

Of course line 10 in Algorithm 2 is highly inefficient. We can improve this strategy for example by rerunning the initial heuristic for the TMP but might find out, that we need to rerun the cleaning process because again a track with more than b cars might occur. But this heuristic gives a good overview about the error of using a strategy to solve TMP for an instance of b-bounded TMP.

Another approach might be to adjust Algorithm 1 so that it stops as soon as the number of b cars per track is exceeded. See Algorithm 3 for pseudocode. Here several adjustments have to be made. Not only do we need to check the length of a track each time destinations are assigned, but we also need to check if there are destinations c with $w(c) > b$. These destinations need to be split to several tracks.

Figures 7 and 8 summarize the output of both heuristics on 50 random instances with a fixed number of 150 cars. The performance of both heuristics depends highly on the ordering of the incoming train and the bound b. Having $b = 10$ the solution of b-bounded TMP and unbounded TMP does not differ significantly. For other random instances with less destinations this might be different. On the other hand we can see that CLEAN-B-bounded is highly inefficient. But it gives a good idea of the error when applying an unbounded approach to the b-bounded TMP.

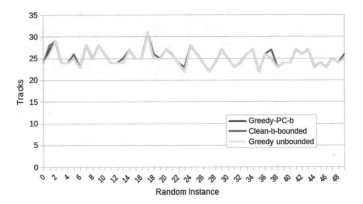

Fig. 8 Results for random instances with $n = 150$ cars solving 10-bounded TMP. We use a u_{greedy} introduced as novel heuristic to solve unbounded TMP to compare the results of b-bounded to unbounded TMP. GREEDY-PC-B-BOUNDED is an adjusted version of this heuristic, CLEAN-B-BOUNDED tries to find a feasible solution by fixing the unbounded solution

10 Conclusions

We have introduced and discussed pseudochain partitions and their relation to the Train Marshalling Problem. There is only little discussion about TMP in the literature, but the problem has an intimate relationship to other sorting problems of rolling stock, see [4]. Thus it is an important step to provide a better understanding of the underlying graph structures. Pseudochain partitions directly lead to a new heuristic providing a improved upper bounds for optimal solutions of TMP. We could proof that every optimal solution of TMP is equivalent to a minimal partition of the corresponding extended comparability graph $D^*(S)$ into pseudochains.

The greedy approach has been evaluated for two sets of instances with 100 and 300 cars, each consisting of 50 random instances each. The computational results show that the model is useful and the proposed Greedy-approach performs in general significantly better than other state-of-the art approaches.

In addition we showed that a weighted version of pseudochains can also be used to solve the b-bounded version of the TMP. We evaluated two simple strategies for this problem.

To sum up, although we achieved encouraging results, there are still questions which are not answered or even discussed in this paper. For example, can the inherent structure of pseudoschains be used to find even better heuristics than those discussed in this paper? Are there any instances of the TMP that can be solved in polynomial time?

The results encourage to look for further improvements to heuristics to solve minPC and to apply the methods to other sorting of rolling Stock Problems.

References

1. Beygang, K., Dahms, F., Krumke, S.O.: Train Marshalling Problem—Algorithms and Bounds. Technical Report, vol. 132. University of Kaiserslautern (2010)
2. Hiller, W.: Rangierbahnhöfe. VEB Verlag für Verkehrswesen, Berlin (1983)
3. Giger, M.: Rangierbahnhof limmattal. In: Swiss Engineering STZ Automate Now!. vol. 1–2, pp. 93–94 (2010)
4. Hansmann, R.S.: Optimal Sorting of Rolling Stock. Cuvillier, Göttingen (2011)
5. Zhu, Y., Zhu, R.: Sequence reconstruction under some order-type constraints. Sci. Sin. (A) **26**(7), 702–713 (1983)
6. Dahlhaus, E., Horak, P., Miller, M., Ryan, J.: The train marshalling problem. Discret. Appl. Math. **103**(1–3), 41–54. https://doi.org/10.1016/s0166-218x(99)00219-x (2000)
7. Brueggeman, L., Fellows, M., Fleischer, R., Lackner, M., Komusiewicz, C., Koutis, Y., Pfandler, A., Rosamond, F.: Train marshalling is fixed parameter tractable. In: Kranakis, E., Krizanc, D., Luccio, F. (eds.) Fun with Algorithms, pp. 51–56. Springer, Berlin, Heidelberg. https://doi.org/10.1007/978-3-642-30347-0_8 (2012)
8. Dahlhaus, E., Manne, F., Miller, M., Ryan, J.: Algorithms for combinatorial problems related to train marshalling. In: Proceedings of AWOCA 2000, pp. 7–16, Hunter Valley (2000)
9. Beygang, K.: On the Solution of Some Railway Freight Car optimization Problems. Dr. Hut, München (2011)
10. Rinaldi, F., Rizzi, R.: Solving the train marshalling problem by inclusion—exclusion. Discret. Appl. Math. **217**)(Part 3), 685–690 (2017)
11. Haahr, J.T., Lusby, R.M.: A matheuristic approach to integrate humping and pullout sequencing operations at railroad hump yards. Networks **67**(2), 126–138 (2016)
12. Dörpinghaus, J., Schrader, R.: A graph-theoretic approach to the train marshalling problem. In: Ganzha, M., Maciaszek, L., Paprzycki, M.(eds.) Proceedings of the 2018 Federated Conference on Computer Science and Information Systems, series Annals of Computer Science and Information Systems, vol. 15, pp. 227–231. IEEE. https://doi.org/10.15439/2018F26 (2018)
13. Dörpinghaus, J.: Pseudostabile Mengen in Graphen. Fraunhofer Verlag. https://kups.ub.uni-koeln.de/8066/ (2018)
14. Bohlin, M., Hansmann, R., Zimmermann, U.T.: Optimization of railway freight shunting. In: Handbook of Optimization in the Railway Industry, pp. 181–212. Springer (2018)

InterCriteria Analysis of Different Hybrid Ant Colony Optimization Algorithms for Workforce Planning

Stefka Fidanova, Olympia Roeva, Gabriel Luque and Marcin Paprzycki

Abstract Every organization and factory optimize their production process with a help of workforce planing. The aim is minimization of the assignment costs of the workers, who will do the jobs. The problem is very complex and needs exponential number of calculations, therefore special algorithms are developed to be solved. The problem is to select employers and to assign them to the jobs to be performed. This problem has very strong constraints and it is difficult to find feasible solutions. The objective is to fulfil the requirements and to minimize the assignment cost. We propose a hybrid Ant Colony Optimization (ACO) algorithm to solve the workforce problem, which is a combination between ACO and an appropriate local search procedure. In this investigation InterCriteria Analysis (ICrA) is applied over numerical results obtained from ACO algorithms with the suggested different variants of local search procedures. Based on ICrA the ACO hybrid algorithms performance is examined and compared.

Keywords Workforce planning · Ant Colony Optimization · Metaheuristics · Local search · InterCriteria analysis · Index matrix · Intuitionistic fuzzy sets

S. Fidanova (✉)
Institute of Information and Communication Technology, Bulgarian Academy of Sciences, Sofia, Bulgaria
e-mail: stefka@parallel.bas.bg

O. Roeva
Institute of Biophysics and Biomedical Engineering, Bulgarian Academy of Sciences, Sofia, Bulgaria
e-mail: olympia@biomed.bas.bg

G. Luque
DLCS, University of Mlaga, 29071 Mlaga, Spain
e-mail: gabriel@lcc.uma.es

M. Paprzycki
System Research Institute Polish Academy of Sciences, Warsaw and Management Academy, Warsaw, Poland
e-mail: marcin.paprzycki@ibspan.waw.pl

© Springer Nature Switzerland AG 2020
S. Fidanova (ed.), *Recent Advances in Computational Optimization*,
Studies in Computational Intelligence 838,
https://doi.org/10.1007/978-3-030-22723-4_5

61

1   Introduction

One of the most important decision making problem, common for all branches of industry, is workforce planning. The workforce planing is a part of the human resource management. It includes multiple level of complexity, therefore it is a hard optimization problem (NP-hard). This problem consists of two decision sets: selection and assignment. The first set shows selected employees from available workers. The assignment set shows which worker which job will perform. The aim is to fulfil the work requirements with minimal assignment cost.

As we mentioned the problem is a hard optimization problem with strong constraints and is impossible to be solved with exact methods or traditional numerical methods for instances with realistic size. These kind of methods can be apply only on some simplified variants of the problem. A deterministic workforce planing problem is studied in [1, 2]. Workforce planing models are reformulated as mixed integer programming in [1]. The authors show that the mixed integer program is much easier to solve than the non-linear program. In [2] the model includes workers differences and the possibility of workers training and upgrading. In [3, 4] a variant with random demands of the problem is proposed. Two stage program of scheduling and allocating with random demands is considered in [3]. Other variant of the problem is to include uncertainty [5–9]. Most of the authors simplify the problem by omitting some of the constraints. In [10] a mixed linear programming is applied and in [4] a decomposition method is applied. For the more complex non-linear workforce planning problems, the convex methods are not applicable.

Nowadays, nature-inspired metaheuristic methods receive great attention [11–15]. In considered here problem some heuristic method including genetic algorithm [16, 17], memetic algorithm [18], scatter search [16] etc., are applied.

So far the Ant Colony Optimization (ACO) algorithm is proved to be very effective solving various complex optimization problems [19, 20]. In our previous work [21] we propose ACO algorithm for workforce planning. We have considered the variant of the workforce planning problem proposed in [16]. Current paper is the continuation of [21]. We propose a hybrid ACO algorithm which is a combination of ACO with a local search procedure. The aim is to improve the algorithm performance.

In order to compare the proposed hybrid ACO algorithm with different local search procedures the approach named InterCriteria Analysis (ICrA) is used. ICrA aiming to go beyond the nature of the criteria involved in a process of evaluation of multiple objects against multiple criteria, and, thus to discover some dependencies between the ICrA criteria themselves [22]. The approach is based on the apparatus of the index matrices and the intuitionistic fuzzy sets, two approaches that have been actively researched and applied [23–28].

For the first time ICrA has been applied for the purposes of temporal, threshold and trends analyses of an economic case-study of European Union member states' competitiveness [29–31]. The approach already has a lot of different applications [32–34]. ICrA is applied for comparison of different metaheuristics as GAs and ACO [35, 36], too.

In this paper ICrA is applied for analysis of an ACO algorithm for workforce planing combined with various local search procedures. The aim is to analyze the algorithm performance according the local search procedures and to study the correlations between the different variants.

The rest of the paper is organized as follows. In Sect. 2 the mathematical description of the problem is presented. In Sect. 3 hybrid ACO algorithm for workforce planing problem is proposed. In Sect. 4 a brief discussion on InterCriteria Analysis background is done. Section 5 shows numerical calculations, ACO algorithms comparisons and discussion of the results. In Sect. 6 results form InterCriteria Analysis application are discussed. A conclusion and directions for future work are done in Sect. 7.

2 Definition of the Workforce Planning Problem

In this paper we solve the workforce planning problem proposed in [16, 37]. The set of jobs $J = \{1, \ldots, m\}$ must be completed during a fixed period of time. The job j requires d_j hours to be completed. $I = \{1, \ldots, n\}$ is the set of workers, candidates to be assigned. Every worker must perform every of assigned to him job minimum h_{min} hours to can work in efficient way. Availability of the worker i is s_i hours. One worker can be assigned to maximum j_{max} jobs. The set A_i shows the jobs, that worker i is qualified. Maximum t workers can be assigned during the planed period, or at most t workers may be selected from the set I of workers. The selected workers need to be capable to complete all the jobs. The aim is to find feasible solution, that optimizes the objective function.

Let c_{ij} is the cost of assigning the worker i to the job j. The mathematical model of the workforce planing problem can be described as follows:

$$x_{ij} = \begin{cases} 1 \text{ if the worker } i \text{ is assigned to job } j \\ 0 \text{ otherwise} \end{cases}$$

$$y_i = \begin{cases} 1 \text{ if worker } i \text{ is selected} \\ 0 \text{ otherwise} \end{cases}$$

$$z_{ij} = \text{number of hours that worker } i$$

$$\text{is assigned to perform job } j$$

$$Q_j = \text{set of workers qualified to perform job } j$$

$$\text{Minimize} \sum_{i \in I} \sum_{j \in A_i} c_{ij} \cdot x_{ij} \tag{1}$$

subject to

$$\sum_{j \in A_i} z_{ij} \leq s_i \cdot y_i \quad i \in I \tag{2}$$

$$\sum_{i \in Q_j} z_{ij} \geq d_j \quad j \in J \tag{3}$$

$$\sum_{j \in A_i} x_{ij} \leq j_{max} \cdot y_j \quad i \in I \tag{4}$$

$$h_{min} \cdot x_{ij} \leq z_{ij} \leq s_i \cdot x_{ij} \quad i \in I, j \in A_i \tag{5}$$

$$\sum_{i \in I} y_i \leq t \tag{6}$$

$$x_{ij} \in \{0, 1\} \ i \in I, j \in A_i$$
$$y_i \in \{0, 1\} \ i \in I$$
$$z_{ij} \geq 0 \qquad i \in I, j \in A_i$$

The objective function is the minimization of the total assignment cost. The number of hours for each selected worker is limited (inequality 2). The work must be done in full (inequality 3). The number of the jobs, that every worker can perform is limited (inequality 4). There is minimal number of hours that every job must be performed by every assigned worker to can work efficiently (inequality 5). The number of assigned workers is limited (inequality 6).

The same model can be used with different objective functions. Minimization of total assignment cost is the aim of this paper. If $\tilde{c}_{ij}$ is the cost the worker i to performs the job j for one hour, than the objective function can minimize the cost of the hall jobs to be finished.

$$f(x) = \text{Min} \sum_{i \in I} \sum_{j \in A_i} \tilde{c}_{ij} \cdot x_{ij} \tag{7}$$

The workforce planning problem is difficult to be solved because of very restrictive constraints especially the relation between the parameters h_{min} and d_j. When the problem is structured (d_j is a multiple of h_{min}), in this case it is more easier to find feasible solution, than for unstructured problems (d_j and h_{min} are not related).

3   Hybrid Ant Colony Optimization Algorithm

The ACO is a nature inspired method. It is metaheuristics methodology following the behaviour of real ants looking for a food. Real ants use chemical substance, called pheromone, to mark their path ant to can return back. An ant moves in random way and when it detects a previously laid pheromone it decides whether to follow it and reinforce it with a new added pheromone. Thus the more ants follow a given trail, the more attractive that trail becomes. Using their collective intelligent the ants can find a shorter path between the source of the food and the nest.

3.1 Main ACO Algorithm

A lot of problems coming from real life and industry needs exponential number of calculations. It is not practical to solve them with exact methods or traditional numerical methods when the problem is large. Therefore the only option is to be applied some metaheuristics. The goal is to find a good solution for a reasonable time [38].

First idea to use ant behaviour for solving optimization problems is proposed by Marco Dorigo [39]. Later some modifications are proposed, mainly in pheromone updating rules [38]. The basic in ACO methodology is the simulation of ants behaviour. The problem is represented by graph. The solutions are represented by paths in a graph and we look for shorter path corresponding to given constraints. The requirements of ACO algorithm are as follows:

- Appropriate representation of the problem by a graph;
- Appropriate pheromone placement on the nodes or on the arcs of the graph;
- Suitable problem-dependent heuristic function, which manage the ants to improve solutions;
- Pheromone updating rules;
- Transition probability rule, which specifies how to include new nodes in the partial solution.

The transition probability $p_{i,j}$, to choose the node j, when the current node is i, is a product of the heuristic information $\eta_{i,j}$ and the pheromone trail level $\tau_{i,j}$ related with this move, where $i, j = 1, \ldots, n$.

$$p_{i,j} = \frac{\tau_{i,j}^a \eta_{i,j}^b}{\sum\limits_{k \in Unused} \tau_{i,k}^a \eta_{i,k}^b}, \quad (8)$$

where $Unused$ is the set of unused nodes of the graph.

A node becomes more profitable if the value of the heuristic information and/or the related pheromone is higher. At the beginning, the initial pheromone level is the same for all elements of the graph and is set to a small positive constant value τ_0, $0 < \tau_0 < 1$. At the end of every iteration the ants update the pheromone values. Different ACO algorithms adopt different criteria to update the pheromone level [38].

The main pheromone trail update rule is:

$$\tau_{i,j} \leftarrow \rho \tau_{i,j} + \Delta \tau_{i,j}, \quad (9)$$

where ρ decreases the value of the pheromone, like the evaporation in a nature. $\Delta \tau_{i,j}$ is a new added pheromone, which is proportional to the quality of the solution. The quality of the solution is measured by the value of the objective function of the solution constructed by the ant.

The starting node for every ant is randomly chosen. It is a diversification of the search. Because the random start a relatively few number of ants can be used, comparing with other population based metaheuristics. The heuristic information represents the prior knowledge of the problem, which we use to better manage the ants. The pheromone is a global experience of the ants to find optimal solution. The pheromone is a tool for concentration of the search around best so far solutions.

3.2 ACO Algorithm for Workforce Planning

In this section we will apply the ACO algorithm for workforce planing from our previous work [21], which is without local search procedure. One of the main points of the ant algorithm is the proper representation of the problem by graph. In our case the graph of the problem is 3 dimensional and the node (i, j, z) corresponds worker with number i to be assigned to the job j for time z. The graph of the problem is asymmetric, because the maximal value of z depends of the value of j, different jobs needs different time to be completed. At the beginning of every iteration every ant starts to construct their solution, from random node of the graph of the problem. For every ant are generated three random numbers. The first random number corresponds to the worker we assign and is in the interval $[0, \ldots, n]$. The second random number corresponds to the job which this worker will perform and is in the interval $[0, \ldots, m]$. We verify if the worker is qualified to perform the job, if not, we chose in a random way another job. The third random number corresponds to the number of hours worker i is assigned to performs the job j and is in the interval $[h_{min}, \ldots, \min\{d_j, s_i\}]$. After, the ant applies the transition probability rule to include next nodes in the partial solution, till the solution is completed, or there is not a possibility to include new node.

We propose the following heuristic information:

$$\eta_{ijl} = \begin{cases} 1/c_{ij} & l = z_{ij} \\ 0 & otherwise \end{cases} \tag{10}$$

This heuristic information stimulates to assign the most cheapest worker as longer as possible. the node with a highest probability is chosen to be the next node, included in the solution. When there are several candidate nodes with a same probability, the next node is chosen between them in a random way.

When a new node is included we take in to account the problem constraints as follows: how many workers are assigned till now; how many time slots every worker is assigned till now; how many time slots are assigned per job till now. When some move of the ant do not meets the problem constraints, then the probability of this move is set to be 0. If for all possible nodes the value of the transition probability is 0, it is impossible to include new node in the solution and the solution construction stops. When the constructed solution is feasible the value of the objective function is the sum of the assignment cost of the assigned workers. If the constructed solution is not feasible, the value of the objective function is set to be equal to -1.

The ants constructed feasible solutions deposed a new pheromone on the elements of their solutions. The new added pheromone is equal to the reciprocal value of the objective function.

$$\Delta \tau_{i,j} = \frac{\rho - 1}{f(x)} \tag{11}$$

Thus the nodes of the graph belonging to solutions with less value of the objective function, receive more pheromone than others and become more desirable in the next iteration.

At the end of every iteration we compare the iteration best solution with the best so far solution. If the best solution from the current iteration is better than the best so far solution (global best solution), we update the global best solution with the current iteration best solution.

The end condition used in our algorithm is the number of iterations.

3.3 Local Search Procedure

The our main contribution in this paper is the hybridization of the ACO algorithm with a local search procedure. The aim of the local search is to decrease the time to find the best solution and eventually to improve the achieved solutions.

We apply local search procedure only on infeasible solutions and only one time disregarding the new solution is feasible or not. Thus, our local search is not time consuming. The workforce planning is a problem with strong constraints and part of the ants do not succeed to find feasible solution. With our local search procedure the possibility to find feasible solution increases and thus increases the chance to improve current solution.

If the solution is not feasible we remove part of the assigned workers and after that we assign in their place new workers. The workers which will be removed are chosen randomly. On this partial solution we assign new workers applying the rules of ant algorithm. The ACO algorithm is a stochastic algorithm, therefore the new constructed solution is different from previous one with a high probability.

We propose three variants of local search procedure:

- removed workers are quarter of all assigned workers;
- removed workers are half of all assigned workers;
- all assigned workers are removed.

4 InterCriteria Analysis

According to [22, 40–42], we will obtain an Intuitionistic Fuzzy Pair (IFP) as the degrees of "agreement" and "disagreement" between two criteria applied on different objects. An IFP is an ordered pair of real non-negative numbers $\langle a, b \rangle$ such that:

$$a + b \leq 1.$$

Let us be given an IM [43] whose index sets for rows consist of the names of the criteria and for columns—objects. We will obtain an IM with index sets consisting of the names of the criteria both for rows and for columns. The elements IFPs of this IM corresponds to the degrees of "agreement" and degrees of "disagreement" of the considered criteria.

The following two points are supposed:

1. All criteria provide an evaluation for all objects and all these evaluations are available.
2. All the evaluations of a given criteria can be compared amongst themselves.

Further, by O we denote the set of all objects $O_1, O_2, \ldots, O_n$ being evaluated, and by $C(O)$ the set of values assigned by a given criteria C to the objects, i.e.

$$O \stackrel{\text{def}}{=} \{O_1, O_2, \ldots, O_n\},$$

$$C(O) \stackrel{\text{def}}{=} \{C(O_1), C(O_2), \ldots, C(O_n)\}.$$

Let $x_i = C(O_i)$. Then the following set can be defined:

$$C^*(O) \stackrel{\text{def}}{=} \{\langle x_i, x_j \rangle | i \neq j \,\&\, \langle x_i, x_j \rangle \in C(O) \times C(O)\}.$$

In order to find the degrees of "agreement" of two criteria the vector of all internal comparisons of each criteria is constructed. This vector fulfil exactly one of the following three relations—R, $\overline{R}$ and $\tilde{R}$. For a fixed criterion C and any ordered pair $\langle x, y \rangle \in C^*(O)$ it is required:

$$\langle x, y \rangle \in R \Leftrightarrow \langle y, x \rangle \in \overline{R} \tag{12}$$

$$\langle x, y \rangle \in \tilde{R} \Leftrightarrow \langle x, y \rangle \notin (R \cup \overline{R}) \tag{13}$$

$$R \cup \overline{R} \cup \tilde{R} = C^*(O) \tag{14}$$

From the above it is seen that We only need to consider a subset of $C(O) \times C(O)$ for the effective calculation of $V(C)$ (vector of internal comparisons). From Eqs. (12)–(14) it follows that if we know what is the relation between x and y, we also know what is the relation between y and x. Thus, we will only consider lexicographically ordered pairs $\langle x, y \rangle$.

Let:
$$C_{i,j} = \langle C(O_i), C(O_j) \rangle.$$

We construct the vector with exactly $\frac{n(n-1)}{2}$ elements:

$$V(C) = \{C_{1,2}, C_{1,3}, \ldots, C_{1,n}, C_{2,3}, C_{2,4}, \ldots, C_{2,n},$$

$$C_{3,4}, \ldots, C_{3,n}, \ldots, C_{n-1,n}\}.$$

for a fixed criterion C.

Further, we replace the vector $V(C)$ with $\hat{V}(C)$, where for each $1 \leq k \leq \frac{n(n-1)}{2}$ for the k-th component it is true:

$$\hat{V}_k(C) = \begin{cases} 1 \text{ iff } V_k(C) \in R, \\ -1 \text{ iff } V_k(C) \in \overline{R}, \\ 0 \text{ otherwise.} \end{cases}$$

We determine the degree of "agreement" ($\mu_{C,C'}$) between the two criteria as the number of matching components. This can be done in several ways, e.g. by counting the matches or by taking the complement of the Hamming distance. The degree of "disagreement" ($\nu_{C,C'}$) is the number of components of opposing signs in the two vectors. This may also be done in various ways.

It is obvious that:

$$\mu_{C,C'} = \mu_{C',C},$$

$$\nu_{C,C'} = \nu_{C',C},$$

and $\langle \mu_{C,C'}, \nu_{C,C'} \rangle$ is an IFP.

The difference

$$\pi_{C,C'} = 1 - \mu_{C,C'} - \nu_{C,C'} \tag{15}$$

is considered as a degree of "uncertainty".

5 Results

5.1 Numerical Results

In this section test results are reported and compared with ACO algorithm without local search procedure. The software which realizes the algorithm is written in C computer language and is run on Pentium desktop computer at 2.8 GHz with 4 GB RAM.

We use the artificially generated problem instances considered in [16]. The test instances characteristics are shown in Table 1.

The set of test problems consists of ten structured and ten unstructured problems. The structured problems are enumerated from $S01$ to $S10$ and unstructured problems are enumerated from $U01$ to $U10$. For structured problems d_j is proportional to h_{min}. In our previous work [21] we show that our ACO algorithm outperforms the genetic and scatter search algorithms from [16].

Table 1 Test instances characteristics

Parameters	Value
n	20
m	20
t	10
s_i	[50, 70]
j_{max}	[3, 5]
h_{min}	[10, 15]

The number of iterations is a stopping criteria for our hybrid ACO algorithm. The number of iterations is fixed to be maximum 100. In Table 2 the parameter settings of our ACO algorithm are shown. The values are fixed experimentally.

Further, the problem instances are enumerated as $S20_{01}$–$S20_{10}$ and $U20_{01}$ to $U20_{10}$, taking into account the number of ants.

The workforce problem has very restrictive constraints. Therefore only 2–3 of the ants, per iteration, find feasible solution. Sometimes exist iterations without any feasible solution. Its complicates the search process. Our aim is to decrease the number of unfeasible solutions and thus to increase the possibility ants to find good solutions and so to decrease needed number of iterations to find good solution. We observe that after the local search procedure applied on the first iteration, the number of unfeasible solutions in a next iterations decrease. It is another reason the computation time does not increase significantly. We are dealing with four cases:

- without local search procedure (ACO);
- local search procedure when the number of removed workers is quarter from the number of all assigned workers (ACO quarter);
- local search procedure when the number of removed workers is half from the number of all assigned workers (ACO half);
- local search procedure when all assigned workers are removed and the solution is constructed from the beginning (ACO restart);

We perform 30 independent runs with every one of the four cases, because the algorithm is stochastic ant to guarantee the robustness of the average results. We apply ANOVA test for statistical analysis to guarantee the significance of the achieved results. The obtained results are presented in Tables 3 and 4. Table 3 presents the minimal number of iterations to achieve the best solution and Table 4—the computation time needed to achieve the best solution.

Table 2 ACO parameter settings

Parameters	Value
Number of iterations	100
ρ	0.5
τ_0	0.5
Number of ants	20
a	1
b	1

Table 3 Minimal number of iterations to achieve the best solution

	ACO	ACO quarter	ACO half	ACO restart
$S20_{01}$	13	10	15	16
$S20_{02}$	17	28	28	35
$S20_{03}$	29	27	37	33
$S20_{04}$	77	66	41	23
$S20_{05}$	21	21	4	14
$S20_{06}$	21	13	20	1
$S20_{07}$	43	34	29	40
$S20_{08}$	57	15	50	33
$S20_{09}$	36	28	22	48
$S20_{10}$	26	19	16	35
$U20_{01}$	17	23	11	21
$U20_{02}$	17	16	12	15
$U20_{03}$	28	22	20	48
$U20_{04}$	41	56	28	28
$U20_{05}$	14	20	15	4
$U20_{06}$	46	46	45	20
$U20_{07}$	29	44	37	39
$U20_{08}$	11	14	16	26
$U20_{09}$	46	68	41	42
$U20_{10}$	30	30	30	30

We are interested of the number of iterations for finding the best result. It can be very different for different test problems, so we will use ranking of the algorithms. The variant of our hybrid algorithm is on the first place, if it achieves the best solution with less average number of iterations over 30 runs, according other cases and we assign to it 1, we assign 2 to the case on the second place, 3 to the case on the third place and 4 to the case with most number of iterations. On some cases can be assigned same numbers if the number of iterations to find the best solution is the same. We sum the ranking of the cases over all 20 test problems to find final ranking of the different cases of the hybrid algorithm.

We observe that the local search procedure decreases the number of unfeasible solutions, found by traditional ACO algorithm in the next iterations, thus when the number of iterations increase, the need of local search procedure decreases. On Table 5 we report the achieved ranking of different cases of our hybrid algorithm. As we mentioned above, with ACO quarter we call the case when quarter of the workers are removed. ACO half is the case when half of the workers are removed. ACO restart is the case when all workers are removed. It is like to restart the solution construction, to construct the solution from the beginning.

Table 4 Computation time needed to achieve the best solution

	ACO	ACO quarter	ACO half	ACO restart
$S20_{01}$	1.20	0.94	0.96	2.29
$S20_{02}$	3.94	8.62	6.22	14.75
$S20_{03}$	5.19	5.79	11.93	3.06
$S20_{04}$	3.06	16.66	7.00	6.11
$S20_{05}$	0.63	1.312	0.396	0.90
$S20_{06}$	2.48	2.12	2.64	0.59
$S20_{07}$	6.78	4.82	6.78	6.60
$S20_{08}$	6.38	1.87	10.42	8.59
$S20_{09}$	4.68	5.31	4.48	5.70
$S20_{10}$	1.45	1.25	1.28	10.43
$U20_{01}$	3.10	4.48	2.00	2.50
$U20_{02}$	1.98	1.18	0.92	0.93
$U20_{03}$	2.14	2.41	1.54	1.88
$U20_{04}$	3.08	3.35	3.12	3.47
$U20_{05}$	1.55	2.76	2.06	1.056
$U20_{06}$	10.92	11.8	4.36	7.05
$U20_{07}$	4.22	6.55	3.54	3.27
$U20_{08}$	0.89	1.48	1.19	1.77
$U20_{09}$	6.48	8.72	7.10	7.21
$U20_{10}$	3.74	3.88	3.69	10.00

One of the main questions is how many worker to remove from unfeasible solution, so that the new constructed solution to be feasible and close to the best one. We calculate the ranking, regarding the average number of iterations to find best solution over 30 runs of the test. When more than half of the ants find unfeasible solutions, the deviation from the average is larger compared to the tests when the most of the ants achieve feasible solutions. Table 5 shows that the local search procedure decreases the number of iterations needed to find the best solution, when more than half of the workers are removed. The traditional ACO algorithm and hybrid ACO with removing quarter of the workers are 4 times on the first place when either, by chance, the algorithm find the best solution on the first iteration, or all ants find feasible solutions. We observe that the both cases are on the third and forth place 12 times. This means that removing less than half of the workers is not enough to construct feasible solution. The ACO algorithm with removed half of the workers 15 times is on the first or second place and only one time is on the fourth place, which means that it performs much better than previous two cases. When all workers are removed the achieved ranking is similar to the case when half of the workers are removed. Let the maximal number of assigned workers is t. Thus the every one

Table 5 Hybrid ACO ranking over number of iterations

	ACO	ACO quarter	ACO half	ACO restart
First place	4 times	4 times	8 times	8 times
Second place	4 times	4 times	7 times	6 times
Third place	8 times	6 times	4 times	3 times
Forth place	4 times	6 times	1 times	3 times
Ranking	52	54	38	41

Table 6 Hybrid ACO comparison over computation time

	ACO	ACO quarter	ACO half	ACO restart
First place	4 times	3 times	10 times	6 times
Second place	7 times	5 times	5 times	5 times
Third place	8 times	4 times	3 times	4 times
Forth place	1 times	8 times	2 times	5 times
Average time (s)	82.244	93.98	79.63	103.012

of the solutions consists about t workers. If all of the workers are removed, the ant need to add new workers on their place which number is about t. When half of the workers are removed, then the ant will add about $t/2$ new workers. The computation time to remove and add about $t/2$ workers is about two times less than to remove and add about t workers. Thus we can conclude that the local search procedure with removing half of the workers performs better than other cases.

Another way for comparison is the computation time. For every test problem and every case we calculate the average time to achieve best solution over 30 runs. In Table IV we did similar ranking as in Table 5, but taking in to account the computation time instead number of iterations. Regarding the Table 6 the ranking according the time is similar to the ranking according to the number of iterations from the Table 5. The best performance is when half of the worker are removed in the local search procedure and the worst performance is when quarter of the workers are removed. The local search procedure with removing half of the worker is on the first place 10 times and on the forth place only 2 times. The local search procedure with removing quarter of the workers is on the first place 3 times and on the forth place 8 times. Regarding the computation time the local search procedure with removing half of the workers again is the best, but the worst is the local search procedure with removing all workers. Reconstructing a solution from the beginning takes more time than to reconstruct partial solution, therefore ACO algorithm with local search procedure removing all workers performs worst. The results from Table 6 show that removing only quarter of the workers from the solution is not enough for construction of good solution and is time consuming comparing with traditional ACO algorithm. According the both types of comparison, ranking and computation time ACO with local search procedure removing half of the assigned workers performs best.

6   Results from InterCriteria Analisys

The four considered here cases of ACO hybrid algorithms will further be denoted as:

- without local search procedure—$ACO1$ (ACO);
- local search procedure when the number of removed workers is quarter from the number of all assigned workers—$ACO2$ (ACO quarter);
- local search procedure when the number of removed workers is half from the number of all assigned workers—$ACO3$ (ACO half);
- local search procedure when all assigned workers are removed and the solution is constructed from the beginning—$ACO4$ (ACO restart).

   The cross-platform software for ICrA approach, ICrAData, is used [44]. The input index matrices for ICrA have the form of Tables 3 and 4. The test problems $S20_{01}$ to $S20_{10}$ and $U20_{01}$ to $U20_{10}$ are considered as objects $O = \{O_1, O_2, \ldots, O_{10}\}$. The four ACO hybrid algorithms ($ACO1 - ACO4$) are considered as criteria $C(O) = \{C(O_1), C(O_2), \ldots, C(O_{10})\}$. In a result, ICrA gives the relation between the proposed ACO hybrid algorithms. The hybrid algorithms are compared based on the obtained results according to the number of iterations and according to the computation time.

6.1   InterCriteria Analisys of Different Hybrid ACO Algorithms According to the Number of Iterations

In Tables 7 and 8 the results of ICrA are presented. To the notation of the hybrid algorithms "it" is added (iterations). The obtained ICrA results are analysed based on the proposed in [45] consonance and dissonance scale. The scheme for defining the consonance and dissonance between each pair of criteria is presented in Table 9.

   The obtained ICrA results are visualized on Fig. 1 within the specific triangular geometrical interpretation of IFSs, thus allowing us to order these results according simultaneously to the degrees of "agreement" $\mu_{C,C'}$ and "disagreement" $\nu_{C,C'}$ of the intuitionistic fuzzy pairs.

   The results show that there is very small values of π, i.e., there is no significant degree of "uncertainty" in the data.

Table 7   Index matrix for $\mu_{C,C'}$ (intuitionistic fuzzy estimations)

$\mu_{C,C'}$	$ACO1_{it}$	$ACO2_{it}$	$ACO3_{it}$	$ACO4_{it}$
$ACO1_{it}$	**1.00**	0.76	0.78	0.60
$ACO2_{it}$	0.76	**1.00**	0.71	0.64
$ACO3_{it}$	0.78	0.71	**1.00**	0.62
$ACO4_{it}$	0.60	0.64	0.62	**1.00**

Table 8 Index matrix for $v_{C,C'}$ (intuitionistic fuzzy estimations)

$v_{C,C'}$	$ACO1_{it}$	$ACO2_{it}$	$ACO3_{it}$	$ACO4_{it}$
$ACO1_{it}$	**0.00**	0.20	0.17	0.35
$ACO2_{it}$	0.20	**0.00**	0.26	0.34
$ACO3_{it}$	0.17	0.26	**0.00**	0.34
$ACO4_{it}$	0.35	0.34	0.34	**0.00**

Table 9 Consonance and dissonance scale [45]

Interval of $\mu_{C,C'}$	Meaning
[0.00–0.05]	Strong negative consonance
(0.05–0.15]	Negative consonance
(0.15–0.25]	Weak negative consonance
(0.25–0.33]	Weak dissonance
(0.33–0.43]	Dissonance
(0.43–0.57]	Strong dissonance
(0.57–0.67]	Dissonance
(0.67–0.75]	Weak dissonance
(0.75–0.85]	Weak positive consonance
(0.85–0.95]	Positive consonance
(0.95–1.00]	Strong positive consonance

Fig. 1 Presentation of ICrA results in the intuitionistic fuzzy interpretation triangle

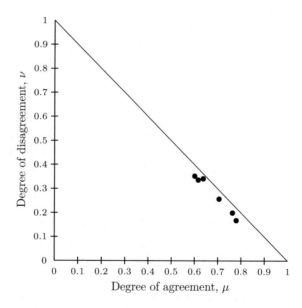

Table 10 ACO hybrid algorithms in weak positive consonance

Pair of ACO hybrid algorithms	$\mu_{C,C'}$-value
$ACO1_{it}$-$ACO3_{it}$	0.78
$ACO1_{it}$-$ACO2_{it}$	0.76

Table 11 ACO hybrid algorithms in weak dissonance

Pair of ACO hybrid algorithms	$\mu_{C,C'}$-value
$ACO2_{it}$-$ACO3_{it}$	0.71

Table 12 ACO hybrid algorithms in dissonance

Pair of ACO hybrid algorithms	$\mu_{C,C'}$-value
$ACO2_{it}$-$ACO4_{it}$	0.64
$ACO3_{it}$-$ACO4_{it}$	0.62
$ACO1_{it}$-$ACO4_{it}$	0.60

Table 13 Index matrix for $\mu_{C,C'}$ (Intuitionistic fuzzy estimations)

$\mu_{C,C'}$	$ACO1_s$	$ACO2_s$	$ACO3_s$	$ACO4_s$
$ACO1_s$	**1.00**	0.77	0.83	0.71
$ACO2_s$	0.77	**1.00**	0.79	0.67
$ACO3_s$	0.83	0.79	**1.00**	0.73
$ACO4_s$	0.71	0.67	0.73	**1.00**

Based on ICrA it is shown that the $ACO1_{it}$-$ACO3_{it}$ and $ACO1_{it}$-$ACO2_{it}$ are the pairs in weak positive consonance (see Table 10). These hybrid algorithms—ACO without local search procedure, ACO quarter and ACO half—show similar performance.

The criteria pair $ACO2_{it}$-$ACO3_{it}$ is in weak dissonance (see Table 11).

Other results show that the pairs $ACO1_{it}$-$ACO4_{it}$, $ACO2_{it}$-$ACO4_{it}$ and $ACO3_{it}$-$ACO4_{it}$ are in strong dissonance, i.e., there is no correlation between them (see Table 12). These results means that the hybrid algorithm ACO restart shows different behaviour compared to the other three ACO algorithms.

6.2 InterCriteria Analisys of Different Hybrid ACO Algorithms According to the Computation Time

Based on the results presented in Table 4 the ICrA is applied. The obtained numerical ICrA results are presented in Tables 13 and 14. To the notation of the hybrid algorithms "s" is added (seconds). The analysis is based again on the proposed in [45] scale (Table 9).

The triangular geometrical interpretation of the obtained results is shown on Fig. 2.

Table 14 Index matrix for $v_{C,C'}$ (Intuitionistic fuzzy estimations)

$v_{C,C'}$	$ACO1_s$	$ACO2_s$	$ACO3_s$	$ACO4_s$
$ACO1_s$	**0.00**	0.23	0.17	0.29
$ACO2_s$	0.23	**0.00**	0.21	0.33
$ACO3_s$	0.17	0.21	**0.00**	0.27
$ACO4_s$	0.29	0.33	0.27	**0.00**

Fig. 2 Presentation of ICrA results in the intuitionistic fuzzy interpretation triangle

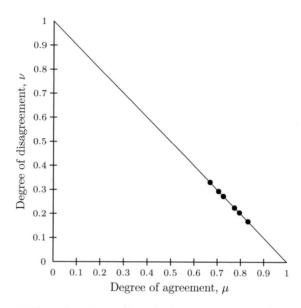

Table 15 ACO hybrid algorithms in weak positive consonance

Pair of ACO hybrid algorithms	$\mu_{C,C'}$-value
$ACO1_s$-$ACO3_s$	0.83
$ACO2_s$-$ACO3_s$	0.79
$ACO1_s$-$ACO2_s$	0.77

In this case $\pi = 0$, i.e., there is no "uncertainty" in the data.

Analysis of the results shows that the pairs of hybrid ACO algorithms e.g. $ACO1_s$-$ACO3_s$ and $ACO1_s$-$ACO2_s$ are in weak positive consonance (see Table 15). We obtain the same results as in the case of comparison based on minimal number of iterations to achieve the best solution. In the case of comparison of computation time needed to achieve the best solution a stronger positive consonance is observed: $\mu_{ACO1_{it},ACO3_{it}} = 0.78$ versus $\mu_{ACO1_s,ACO3_s} = 0.83$ and $\mu_{ACO1_{it},ACO2_{it}} = 0.76$ versus $\mu_{ACO1_s,ACO2_s} = 0.77$. Moreover, the criteria pair $ACO2_s$-$ACO3_s$ that in case of comparison based on minimal number of iterations to achieve the best solution shows weak dissonance, in this case shows weak positive consonance with $\mu_{ACO2_{it},ACO3_{it}} = 0.79$ (Table 15).

Table 16 ACO hybrid algorithms in weak dissonance

Pair of ACO hybrid algorithms	$\mu_{C,C'}$-value
$ACO3_s$-$ACO4_s$	0.73
$ACO1_s$-$ACO4_s$	0.71

Table 17 ACO hybrid algorithms in dissonance

Pair of ACO hybrid algorithms	$\mu_{C,C'}$-value
$ACO2_s$-$ACO4_s$	0.67

Analogically, two criteria pairs, $ACO4_s$-$ACO3_s$ and $ACO1_s$-$ACO4_s$, in this case have moved from dissonance to weak dissonance (Table 16). The criteria pair $ACO2_s$-$ACO4_s$ is in strong dissonance, i.e., there is no correlation between hybrid algorithms ACO quarter and ACO restart (see Table 17).

The presented ICrA results show that there is a difference when the hybrid algorithms are compared based on minimal number of iterations and based on computation time. The comparison based on the computation time is more realistic. In this case, it is compared not only the time to achieve the best solution to compare, but also the time for run of one iteration.

7 Conclusion

In this paper we propose Hybrid ACO algorithm for solving workforce assignment problem. The ACO algorithm is combined with appropriate local search procedure. The local search procedure is applied only on unfeasible solutions. The main idea is to remove part of the workers in the solution in a random way and to include new workers in their place. Three variants of the local search procedure are compared with traditional ACO algorithm, removing quarter of the assigned workers, removing half of the assigned workers and removing all assigned workers. The local search procedure with removing half of the assigned workers performs better than other algorithms.

Further InterCriteria analysis is performed over obtained numerical results solving workforce assignment problem. Additional comparison of the proposed ACO hybrid algorithms is done. Two cases are considered – comparison based on minimal number of iterations to achieve the best solution and based on the computation time needed to achieve the best solution. ICrA results confirm the previously discussed ACO hybrid algorithms performance. Moreover, it is shown that the comparison based on the computation time needed to achieve the best solution is more realistic than the comparison based on minimal number of iterations.

Acknowledgements Work presented here is partially supported by the National Science Fund of Bulgaria under grants DFNI-DN02/10 "New Instruments for Knowledge Discovery from Data and by the Polish-Bulgarian collaborative grant "Practical aspects for scientific computing".

References

1. Hewitt, M., Chacosky, A., Grasman, S., Thomas, B.: Integer programming techniques for solving non-linear workforce planning models with learning. Eur. J. Oper. Res. **242**(3), 942–950 (2015)
2. Othman, M., Bhuiyan, N., Gouw, G.: Integrating workers' differences into workforce planning. Comput. Ind. Eng. **63**(4), 1096–1106 (2012)
3. Campbell, G.: A two-stage stochastic program for scheduling and allocating cross-trained workers. J. Oper. Res. Soc. **62**(6), 1038–1047 (2011)
4. Parisio, A., Jones, C.N.: A two-stage stochastic programming approach to employee scheduling in retail outlets with uncertain demand. Omega (Elsevier) **53**, 97–103 (2015)
5. Hu, K., Zhang, X., Gen, M., Jo, J.: A new model for single machine scheduling with uncertain processing time. J. Intell. Manuf. (Springer) **28**(3), 717–725 (2015)
6. Li, R., Liu, G.: An uncertain goal programming model for machine scheduling problem. J. Intell Manuf. (Springer) **28**(3), 689–694 (2014)
7. Ning, Y., Liu, J., Yan, L.: Uncertain aggregate production planning. Soft Comput. (Springer) **17**(4), 617–624 (2013)
8. Yang, G., Tang, W., Zhao, R.: An uncertain workforce planning problem with job satisfaction. Int. J. Mach. Learn. Cybern. (Springer) (2016). https://doi.org/10.1007/s13042-016-0539-6, http://rd.springer.com/article/10.1007/s13042-016-0539-6
9. Zhou, C., Tang, W., Zhao, R.: An uncertain search model for recruitment problem with enterprise performance. J. Intell. Manuf. (Springer) **28**(3), 295–704 (2014). https://doi.org/10.1007/s10845-014-0997-1
10. Easton, F.: Service completion estimates for cross-trained workforce schedules under uncertain attendance and demand. Prod. Oper. Manag. **23**(4), 660–675 (2014)
11. Albayrak, G., Özdemir, İ.: A state of art review on metaheuristic methods in time-cost trade-off problems. Int. J. Struct. Civil Eng. Res. **6**(1), 30–34 (2017)
12. Mucherino, A., Fidanova, S., Ganzha, M.: Introducing the environment in ant colony optimization, recent advances in computational optimization, studies in computational. Intelligence **655**, 147–158 (2016)
13. Roeva, O., Atanassova, V.: Cuckoo search algorithm for model parameter identification. Int. J. Bioautomation **20**(4), 483–492 (2016)
14. Tilahun, S.L., Ngnotchouye, J.M.T.: Firefly algorithm for discrete optimization problems: a survey. J. Civil Eng. **21**(2), 535–545 (2017)
15. Toimil, D., Gmes, A.: Review of metaheuristics applied to heat exchanger network design. Int. Trans. Oper. Res. **24**(1–2), 7–26 (2017)
16. Alba, E., Luque, G., Luna, F.: Parallel metaheuristics for workforce planning. J. Math. Model. Algorithms (Springer) **6**(3), 509–528 (2007)
17. Li, G., Jiang, H., He, T.: A genetic algorithm-based decomposition approach to solve an integrated equipment-workforce-service planning problem. Omega (Elsevier) **50**, 1–17 (2015)
18. Soukour, A., Devendeville, L., Lucet, C., Moukrim, A.: A memetic algorithm for staff scheduling problem in airport security service. Expert Syst. Appl. **40**(18), 7504–7512 (2013)
19. Fidanova, S., Roeva, O., Paprzycki, M., Gepner, P.: InterCriteria analysis of ACO start startegies. In: Proceedings of the 2016 Federated Conference on Computer Science and Information Systems, pp. 547–550 (2016)
20. Grzybowska, K., Kovcs, G.: Sustainable supply chain—supporting tools. In: Proceedings of the 2014 Federated Conference on Computer Science and Information Systems, vol. 2, pp. 1321–1329 (2014)

21. Fidanova, S., Luquq, G., Roeva, O., Paprzycki, M., Gepner, P.: Ant colony optimization algorithm for workforce planning. In: FedCSIS'2017, IEEE Xplorer, IEEE Catalog Number CFP1585N-ART, pp. 415–419 (2017)
22. Atanassov, K., Mavrov, D., Atanassova, V.: Intercriteria decision making: a new approach for multicriteria decision making, based on index matrices and intuitionistic fuzzy sets. Issues in IFSs and GNs **11**, 1–8 (2014)
23. Traneva, V., Atanassova, V., Tranev, S.: Index matrices as a decision-making tool for job appointment. In: G. Nikolov et al. (eds.) NMA 2018, LNCS , vol. 11189, pp. 1–9. Springer Nature Switzerland AG (2019)
24. Traneva, V., Tranev, S., Atanassova, V.: An intuitionistic fuzzy approach to the hungarian algorithm. In: Nikolov G. et al. (eds.) NMA 2018, LNCS, vol. 11189, pp. 1–9. Springer Nature Switzerland AG (2019)
25. Atanassov, K.T., Vassilev, P.: On the intuitionistic fuzzy sets of n-th type. In: Gaweda A., Kacprzyk J., Rutkowski L., Yen G. (eds.) Advances in Data Analysis with Computational Intelligence Methods. Studies in Computational Intelligence, vol. 738, pp. 265–274. Springer, Cham (2018)
26. Vassilev, P., Ribagin, S.: A note on intuitionistic fuzzy modal-like operators generated by power mean. In: Kacprzyk J., Szmidt E., Zadrony S., Atanassov K., Krawczak M. (eds.) Advances in Fuzzy Logic and Technology 2017. EUSFLAT 2017, IWIFSGN 2017. Advances in Intelligent Systems and Computing, vol. 643, pp. 470–475. Springer, Cham (2018)
27. Marinov, E., Vassilev, P., Atanassov, K.: On separability of intuitionistic fuzzy sets. In: Novel Developments in Uncertainty Representation and Processing, Advances in Intelligent Systems and Computing, vol. 401, pp. 111–123. Springer, Cham (2106)
28. Vassilev, P.: A note on new distances between intuitionistic fuzzy sets. Notes Intuit. Fuzzy Sets **21**(5), 11–15 (2015)
29. Atanassova, V., Mavrov, D., Doukovska, L., Atanassov, K.: Discussion on the threshold values in the intercriteria decision making approach. Notes on Intuit. Fuzzy Sets **20**(2), 94–99 (2014)
30. Atanassova, V., Doukovska, L., Atanassov, K., Mavrov, D.: Intercriteria decision making approach to EU member states competitiveness analysis. In: Shishkov, B. (ed.) Proceedings of the International Symposium on Business Modeling and Software Design—BMSD'14, pp. 289–294 (2014)
31. Atanassova, V. Doukovska, L., Karastoyanov, D., Capkovic, F.: InterCriteria decision making approach to EU member states competitiveness analysis: trend analysis. In: Angelov P. et al. (eds.) Intelligent Systems'2014, Advances in Intelligent Systems and Computing, vol. 322, pp. 107–115 (2014)
32. Roeva, O., Fidanova, S., Vassilev, P., Gepner, P.: InterCriteria analysis of a model parameters identification using genetic algorithm. In: Proceedings of the Federated Conference on Computer Science and Information Systems, vol. 5, pp. 501–506 (2015)
33. Todinova, S., Mavrov, D., Krumova, S., Marinov, P., Atanassova, V., Atanassov, K., Taneva, S.G.: Blood plasma thermograms dataset analysisby means of intercriteria and correlation analyses for the case of colorectal cancer. Int. J. Bioautomation **20**(1), 115–124 (2016)
34. Vassilev, P., Todorova, L., Andonov, V.: An auxiliary technique for intercriteria analysis via a three dimensional index matrix. Notes on Intuit. Fuzzy Sets **21**(2), 71–76 (2015)
35. Angelova, M., Roeva, O., Pencheva, T.: InterCriteria analysis of crossover and mutation rates relations in simple genetic algorithm. In: Proceedings of the Federated Conference on Computer Science and Information Systems, vol. 5, pp. 419–424 (2015)
36. Roeva, O., Fidanova, S., Paprzycki, M.: InterCriteria analysis of ACO and GA hybrid algorithms. Stud. Comput. Intell. **610**, 107–126 (2016)
37. Glover, F., Kochenberger, G., Laguna, M., Wubbena, T.: Selection and assignment of a skilled workforce to meet job requirements in a fixed planning period. In: MAEB04, pp. 636–641 (2004)
38. Dorigo, M., Stutzle, T.: Ant Colony Optimization. MIT Press (2004)
39. Bonabeau, E., Dorigo, M., Theraulaz, G.: Swarm Intelligence: from Natural to Artificial Systems. Oxford University Press, New York (1999)

40. Atanassov, K.: On Intuitionistic Fuzzy Sets Theory. Springer, Berlin (2012)
41. Atanassov, K.: Review and new results on intuitionistic fuzzy sets, mathematical foundations of artificial intelligence seminar, sofia, 1988. Preprint IM-MFAIS-1-88, Reprinted: Int. J. Bioautomation **20**(S1), S7–S16 (2016)
42. Atanassov, K.: Intuitionistic Fuzzy Sets, VII ITKR Session, Sofia, 20–23 June 1983. Reprinted: Int. J. Bioautomation **20**(S1), S1–S6 (2016)
43. Atanassov, K.: On index matrices, Part 1: standard cases. Adv. Stud. Contemp. Math. **20**(2), 291–302 (2010)
44. Ikonomov, N., Vassilev, P., Roeva, O.: ICrAData—software for intercriteria analysis. Int. J. Bioautomation **22**(1), 1–10 (2018)
45. Atanassov, K., Atanassova, V., Gluhchev, G.: InterCriteria analysis: ideas and problems. Notes Intuit. Fuzzy Sets **21**(1), 81–88 (2015)

Different InterCriteria Analysis of Variants of ACO algorithm for Wireless Sensor Network Positioning

Olympia Roeva and Stefka Fidanova

Abstract Wireless sensor networks are formed by spatially distributed sensors, which communicate in a wireless way. This network can monitor various kinds of environment and physical conditions like movement, noise, light, humidity, images, chemical substances etc. A given area needs to be fully covered with minimal number of sensors and the energy consumption of the network needs to be minimal too. We propose several algorithms, based on Ant Colony Optimization, to solve the problem. We study the algorithms behavior when the number of ants varies from 1 to 10. We apply InterCriteria analysis to study relations between proposed algorithms and number of ants and analyse correlation between them. Four different algorithms of ICrA—μ-biased, Balanced, ν-biased and Unbiased—are applied. The obtained results are discussed in order to find the stronger correlations between considered hybrid ACO algorithms.

Keywords Ant colony optimization · InterCriteria analisys · Wireless sensor network

1 Introduction

Wireless Sensor Networks (WSN) allow the monitoring of large areas without the intervention of a human operator. The WSN can be used in areas where traditional networks fail or are inadequate. They find applications in a variety of areas such as climate monitoring, military use, industry and sensing information from inhospitable

O. Roeva
Institute of Biophysics and Biomedical Engineering, Bulgarian Academy
of Sciences, Acad. G. Bonchev Str., bl. 105, 1113 Sofia, Bulgaria
e-mail: olympia@biomed.bas.bg

S. Fidanova (✉)
Institute of Information and Communication Technologies, Bulgarian Academy
of Sciences, Acad. G. Bonchev Str., bl. 25A, 1113 Sofia, Bulgaria
e-mail: stefka@parallel.bas.bg

© Springer Nature Switzerland AG 2020
S. Fidanova (ed.), *Recent Advances in Computational Optimization*,
Studies in Computational Intelligence 838,
https://doi.org/10.1007/978-3-030-22723-4_6

locations. Unlike other networks, sensor networks depend on deployment of sensors over a physical location to fulfill a desired task.

A WSN node contains several components including the radio, battery, micro-controller, analog circuit, and sensor interface. In battery-powered systems, higher data rates and more frequent radio use consume more power. There are several open issues for sensor networks such as signal processing [26], deployment [34], operating cost, localization and location estimation. The wireless sensors, have two fundamental functions: sensing and communicating. However, the sensors which are fare from the high energy communication node can not communicate with him directly. The sensors transmit their data to this node, either directly or via hops, using nearby sensors as communication relays.

Jourdan [19] solved an instance of WSN layout using a multi-objective genetic algorithm—a fixed number of sensors had to be placed in order to maximize the coverage. In some applications most important is the network energy. In [17] is proposed Ant Colony Optimization (ACO) algorithm and in [33] is proposed evolutionary algorithm for this variant of the problem. In [12] is proposed ACO algorithm taking in to account only the number of the sensors. In [13] a multi-objective ACO algorithm, which solves the WSN layout problem is proposed. The problem is multi-objective with two objective functions—(i) minimizing the energy consumption of the nodes in the network, and (ii) minimizing the number of the nodes. The full coverage of the network and connectivity are considered as constraints. A mono-objective ant algorithm which solves the WSN layout problem is proposed in [14]. In [22] are proposed several evolutionary algorithms to solve the problem. In [20] is proposed genetic algorithm which achieves similar solutions as the algorithms in [22], but it is tested on small test problems.

The current research is an attempt to investigate the influence of the number of ants on the ACO algorithm performance, which solves the WSN layout problem, and quality of the achieved solutions and to find the minimal number of ants which are enough to achieve good solutions. For this purpose the InterCriteria Analysis (ICrA) approach is applied.

ICrA, proposed by [10], is a recently developed approach for evaluation of multiple objects against multiple criteria and thus discovering existing correlations between the criteria themselves. It is based on the apparatus of the index matrices (IMs) [4], and the intuitionistic fuzzy sets [7, 8] and can be applied to decision making in different areas of knowledge [6, 23, 24, 28–32]. Various applications of the ICrA approach have been found in science and practice—e-learning [21], algorithms performance [16], medicine [27], etc.

In [25] four different algorithms performing ICrA, namely μ-biased, Balanced, ν-biased and Unbiased, are proposed. The work is based on the [9] where several rules for defining the ways of estimating the degrees of "agreement" and "disagreement", with respect to the type of data, are defined. In this paper the influence of the number of ants on the ACO algorithm performance is investigated based on the above mentioned four ICrA algorithms. Data from series of ACO optimization procedures, published in [13–15], are used to construct IMs. ICrA is applied over the so defined IMs and the results are discussed.

The paper is organized as follows: in Sect. 2 the WSN layout problem formulation is given. In Sect. 3 the background of the InterCriteria Analysis is presented. In Sect. 4 the numerical results are presented and a discussions is provided. The concluding remarks are given in Sect. 5.

2 Problem Formulation

Each sensor node of WSN sense an area around itself called its sensing area, which is determined by the sensing radius. The communication radius determines how far the node can send his data. A special node in the WSN called High Energy Communication Node (HECN) is responsible for external access to the network. Every sensor node in the network must have communication with the HECN, connectivity of the network. The communication radius is often much smaller than the network size, they transmit their date by other nodes which are closer to the HECN.

In our formulation, sensor nodes has to be placed in a terrain providing full sensing coverage with a minimal number of sensors and minimizing the energy spent in communications by any single node. Minimal number of sensors means cheapest network for constructing. Minimal energy means cheapest network for exploitation. The energy of the network defines the lifetime of the network, how frequently the batteries need to be replaced. These are opposed objectives and we look for a good balance between number of sensors and energy consumption.

The WSN operates by rounds: In a round, every node collects the data and sends it to the HECN. Every node transmits the information packets to the neighbor that is closest to the HECN. If a node has n neighbors, each one receives 1/n of its traffic load. Every node has a traffic load equal to 1 (corresponding to its own sent data) plus the sum of all traffic loads received from neighbors.

3 InterCriteria Analysis

InterCriteria analysis, based on the apparatuses of Index Matrices (IM) [1–3] and Intuitionistic Fuzzy Sets (IFS) [5, 8], is given in details in [10]. Here, for completeness, the proposed idea is briefly presented.

Let the initial IM is presented in the form of Eq. (1), where, for every p, q, $(1 \leq p \leq m, 1 \leq q \leq n)$, C_p is a criterion, taking part in the evaluation; O_q—an object to be evaluated; $C_p(O_q)$—a real number (the value assigned by the p-th criteria to the q-th object).

$$A = \begin{array}{c|ccccc} & O_1 & \cdots & O_q & \cdots & O_n \\ \hline C_1 & C_1(O_1) & \cdots & C_1(O_q) & \cdots & C_1(O_n) \\ \vdots & \vdots & \ddots & \vdots & \ddots & \vdots \\ C_p & C_p(O_1) & \cdots & C_p(O_q) & \cdots & C_p(O_n) \\ \vdots & \vdots & \ddots & \vdots & \ddots & \vdots \\ C_m & C_m(O_1) & \cdots & C_m(O_q) & \cdots & C_m(O_n) \end{array} \qquad (1)$$

Let O denotes the set of all objects being evaluated, and $C(O)$ is the set of values assigned by a given criteria C (i.e., $C = C_p$ for some fixed p) to the objects, i.e.,

$$O \stackrel{\text{def}}{=} \{O_1, O_2, O_3, \ldots, O_n\},$$
$$C(O) \stackrel{\text{def}}{=} \{C(O_1), C(O_2), C(O_3), \ldots, C(O_n)\}.$$

Let $x_i = C(O_i)$. Then the following set can be defined:

$$C^*(O) \stackrel{\text{def}}{=} \{\langle x_i, x_j \rangle | i \neq j \ \& \ \langle x_i, x_j \rangle \in C(O) \times C(O)\}.$$

Further, if $x = C(O_i)$ and $y = C(O_j)$, $x \prec y$ will be written iff $i < j$.

In order to find the agreement between two criteria, the vectors of all internal comparisons for each criterion are constructed, which elements fulfill one of the three relations R, $\overline{R}$ and $\tilde{R}$. The nature of the relations is chosen such that for a fixed criterion C and any ordered pair $\langle x, y \rangle \in C^*(O)$:

$$\langle x, y \rangle \in R \Leftrightarrow \langle y, x \rangle \in \overline{R}, \qquad (2)$$
$$\langle x, y \rangle \in \tilde{R} \Leftrightarrow \langle x, y \rangle \notin (R \cup \overline{R}), \qquad (3)$$
$$R \cup \overline{R} \cup \tilde{R} = C^*(O). \qquad (4)$$

Remark 1 For example, if "R" is the relation "$<$", then $\overline{R}$ is the relation "$>$", and vice versa.

For the effective calculation of the vector of internal comparisons (denoted further by $V(C)$) only the subset of $C(O) \times C(O)$ needs to be considered, namely:

$$C^{\prec}(O) \stackrel{\text{def}}{=} \{\langle x, y \rangle | x \prec y \ \& \ \langle x, y \rangle \in C(O) \times C(O),$$

due to Eqs. (2)–(4). For brevity, $c^{i,j} = \langle C(O_i), C(O_j) \rangle$.

Then for a fixed criterion C the vector of lexicographically ordered pair elements is constructed:

$$V(C) = \{c^{1,2}, c^{1,3}, \ldots, c^{1,n}, c^{2,3}, c^{2,4}, \ldots, c^{2,n},$$
$$c^{3,4}, \ldots, c^{3,n}, \ldots, c^{n-1,n}\}. \tag{5}$$

In order to be more suitable for calculations, $V(C)$ is replaced by $\hat{V}(C)$, where its k-th component ($1 \leq k \leq \frac{n(n-1)}{2}$) is given by:

$$\hat{V}_k(C) = \begin{cases} 1, & \text{iff } V_k(C) \in R, \\ -1, & \text{iff } V_k(C) \in \overline{R}, \\ 0, & \text{otherwise.} \end{cases}$$

When comparing two criteria the degree of "agreement" ($\mu_{C,C'}$) is usually determined as the number of matching components of the respective vectors. The degree of "disagreement" ($\nu_{C,C'}$) is usually the number of components of opposing signs in the two vectors. From the way of computation it is obvious that $\mu_{C,C'} = \mu_{C',C}$ and $\nu_{C,C'} = \nu_{C',C}$. Moreover, $\langle \mu_{C,C'}, \nu_{C,C'} \rangle$ is an Intuitionistic Fuzzy Pair (IFP).

There may be some pairs $\langle \mu_{C,C'}, \nu_{C,C'} \rangle$, for which the sum $\mu_{C,C'} + \nu_{C,C'}$ is less than 1. The difference

$$\pi_{C,C'} = 1 - \mu_{C,C'} - \nu_{C,C'} \tag{6}$$

is considered as a degree of "uncertainty".

In this investigation four different algorithms for calculation of $\mu_{C,C'}$ and $\nu_{C,C'}$ are used [25]. A brief description of the ICrA algorithms is presented below:

– **μ-biased ICrA algorithm**: This algorithm follows the rules presented in [9, Table 3], where the rule for =, = for two criteria C and C' is assigned to $\mu_{C,C'}$. An example pseudo-code is presented below as **Algorithm 1**.
– **ν-biased ICrA algorithm**: In this case the rule for =, = for two criteria C and C' is assigned to $\nu_{C,C'}$. It should be noted that in such case a criteria compared to itself does not necessarily yield $\langle 1, 0 \rangle$. An example pseudo-code is presented below as **Algorithm 2**.
– **Balanced ICrA algorithm**: This algorithm follows the rules in [9, Table 2], where the rule for =, = for two criteria C and C' is assigned a half to both $\mu_{C,C'}$ and $\nu_{C,C'}$. It should be noted that in such case a criteria compared to itself does not necessarily yield $\langle 1, 0 \rangle$. An example pseudo-code is presented below as **Algorithm 3**.
– **Unbiased ICrA algorithm**: This algorithm follows the rules in [9, Table 1]. It should be noted that in such case a criterion compared to itself does not necessarily yield $\langle 1, 0 \rangle$, too. An example pseudo-code is presented below as **Algorithm 4**.

Algorithm 1 : *μ-biased*

Require: Vectors $\hat{V}(C)$ and $\hat{V}(C')$

1: **function** DEGREES OF AGREEMENT AND DISAGREEMENT($\hat{V}(C)$, $\hat{V}(C')$)
2:     $V \leftarrow \hat{V}(C) - \hat{V}(C')$
3:     $\mu \leftarrow 0$
4:     $\nu \leftarrow 0$
5:     **for** $i \leftarrow 1$ to $\frac{n(n-1)}{2}$ **do**
6:         **if** $V_i = 0$ **then**
7:             $\mu \leftarrow \mu + 1$
8:         **else if** abs(V_i) = 2 **then**   ▷ abs(V_i): the absolute value of V_i
9:             $\nu \leftarrow \nu + 1$
10:         **end if**
11:     **end for**
12:     $\mu \leftarrow \frac{2}{n(n-1)}\mu$
13:     $\nu \leftarrow \frac{2}{n(n-1)}\nu$
14:     **return** μ, ν
15: **end function**

end

Algorithm 2 : *ν-biased*

Require: Vectors $\hat{V}(C)$ and $\hat{V}(C')$

1: **function** DEGREES OF AGREEMENT AND DISAGREEMENT($\hat{V}(C)$, $\hat{V}(C')$)
2:     $P \leftarrow \hat{V}(C) \odot \hat{V}(C')$ ▷ $\odot$ denotes Hadamard (entrywise) product
3:     $V \leftarrow \hat{V}(C) - \hat{V}(C')$
4:     $\mu \leftarrow 0$
5:     $\nu \leftarrow 0$
6:     **for** $i \leftarrow 1$ to $\frac{n(n-1)}{2}$ **do**
7:         **if** $V_i = 0$ and $P_i \neq 0$ **then**
8:             $\mu \leftarrow \mu + 1$
9:         **else if** $V_i = P_i$ or abs(V_i) = 2 **then**   ▷ abs(V_i): the absolute value of V_i
10:             $\nu \leftarrow \nu + 1$
11:         **end if**
12:     **end for**
13:     $\mu \leftarrow \frac{2}{n(n-1)}\mu$
14:     $\nu \leftarrow \frac{2}{n(n-1)}\nu$
15:     **return** μ, ν
16: **end function**

end

Algorithm 3 *: Balanced*
  Require: Vectors $\hat{V}(C)$ and $\hat{V}(C')$

1: **function** DEGREES OF AGREEMENT AND DISAGREEMENT($\hat{V}(C)$, $\hat{V}(C')$)
2:    $P \leftarrow \hat{V}(C) \odot \hat{V}(C')$ ▷ $\odot$ denotes Hadamard (entrywise) product
3:    $V \leftarrow \hat{V}(C) - \hat{V}(C')$
4:    $\mu \leftarrow 0$
5:    $\nu \leftarrow 0$
6:    **for** $i \leftarrow 1$ to $\frac{n(n-1)}{2}$ **do**
7:      **if** $V_i = P_i$ **then**
8:        $\mu \leftarrow \mu + \frac{1}{2}$
9:        $\nu \leftarrow \nu + \frac{1}{2}$
10:      **else if** $V_i = 0$ and $P_i \neq 0$ **then**
11:        $\mu \leftarrow \mu + 1$
12:      **else if** abs(V_i) = 2 **then**   ▷ abs(V_i): the absolute value of V_i
13:        $\nu \leftarrow \nu + 1$
14:      **end if**
15:    **end for**
16:    $\mu \leftarrow \frac{2}{n(n-1)}\mu$
17:    $\nu \leftarrow \frac{2}{n(n-1)}\nu$
18:    **return** μ, ν
19: **end function**

end

Algorithm 4 *: Unbiased*
  Require: Vectors $\hat{V}(C)$ and $\hat{V}(C')$

1: **function** DEGREES OF AGREEMENT AND DISAGREEMENT($\hat{V}(C)$, $\hat{V}(C')$)
2:    $P \leftarrow \hat{V}(C) \odot \hat{V}(C')$ ▷ $\odot$ denotes Hadamard (entrywise) product
3:    $V \leftarrow \hat{V}(C) - \hat{V}(C')$
4:    $\mu \leftarrow 0$
5:    $\nu \leftarrow 0$
6:    **for** $i \leftarrow 1$ to $\frac{n(n-1)}{2}$ **do**
7:      **if** $V_i = 0$ and $P_i \neq 0$ **then**
8:        $\mu \leftarrow \mu + 1$
9:      **else if** abs(V_i) = 2 **then**   ▷ abs(V_i): the absolute value of V_i
10:        $\nu \leftarrow \nu + 1$
11:      **end if**
12:    **end for**
13:    $\mu \leftarrow \frac{2}{n(n-1)}\mu$
14:    $\nu \leftarrow \frac{2}{n(n-1)}\nu$
15:    **return** μ, ν
16: **end function**

end

As a result of applying any of the proposed algorithms to IM A (Eq. (1)), the following IM is constructed:

$$
\begin{array}{c|ccc}
 & C_2 & \cdots & C_m \\
\hline
C_1 & \langle \mu_{C_1,C_2}, \nu_{C_1,C_2} \rangle & \cdots & \langle \mu_{C_1,C_m}, \nu_{C_1,C_m} \rangle \\
\vdots & \vdots & \ddots & \vdots \\
C_{m-1} & & \cdots & \langle \mu_{C_{m-1},C_m}, \nu_{C_{m-1},C_m} \rangle
\end{array}
,
$$

that determines the degrees of "agreement" and "disagreement" between criteria $C_1, ..., C_m$.

4   Numerical Results and Discussion

In our previous works [12–15] we propose different variants of ACO algorithm to solve WSN problem. In [15] is applied multi-objective ACO algorithm. In [12] the problem is converted to mono-objective by multiplying the two objective functions, and in [14] the problem is converted to mono-objective by summing the two objective functions. We apply our algorithms on rectangular area consisting 500×500 points and the communication and coverage radius of every sensor cover 30 points. The number of used ants is from 1 to 10. For our current research we use results published in these papers.

The input IM and the full set of obtained numerical results could be found in http://intercriteria.net/studies/aco/.

ICA is applied over the data presented in the input IM, where different ACO algorithms (*ACOu*—mono-objective with multiplication, *ACOs*—mono-objective with sum and *ACOm*—multi-objective) are presented as criteria and number of sensors [223, 224, 225, 226, 227, 228, 229, 230, 231, 232, 233, 234, 235, 236, 237, 238, 239, 240, 241, 242, 243, 244]—as objects. The cross-platform software for ICrA approach, ICrAData, is used [18].

The obtained ICrA results are analyzed based on the proposed in [9] consonance and dissonance scale. For ease of use the scheme for defining the consonance and dissonance between each pair of criteria is here presented in Table 1.

The obtained most significant and interesting ICrA results are presented in tables below.

4.1   Results from Application of μ-biased ICrA

All results, based on application of μ-biased ICrA, are visualized on Fig. 1 within the specific triangular geometrical interpretation of IFSs, thus allowing us to order these results according simultaneously to the degrees of "agreement" $\mu_{C,C'}$ and "disagreement" $\nu_{C,C'}$ of the intuitionistic fuzzy pairs.

Table 1 Consonance and dissonance scale [9]

Interval of $\mu_{C,C'}$	Meaning
[0.00–0.05]	Strong negative consonance (SNC)
(0.05–0.15]	Negative consonance (NC)
(0.15–0.25]	Weak negative consonance (WNC)
(0.25–0.33]	Weak dissonance (WD)
(0.33–0.43]	Dissonance (D)
(0.43–0.57]	Strong dissonance (SD)
(0.57–0.67]	Dissonance (D)
(0.67–0.75]	Weak dissonance (WD)
(0.75–0.85]	Weak positive consonance (WPC)
(0.85–0.95]	Positive consonance (PC)
(0.95–1.00]	Strong positive consonance (SPC)

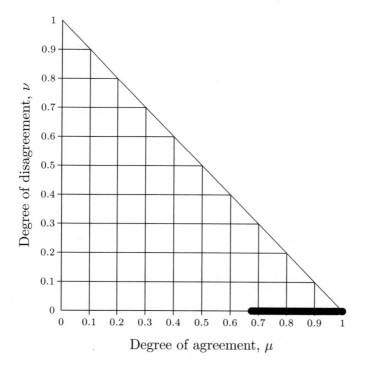

Fig. 1 Presentation of ICrA results in the intuitionistic fuzzy interpretation triangle—μ-biased ICrA

As it can be seen from Fig. 1 the ν-values for all considered pairs are zero ($\nu_{C,C'} = 0$). There are some results with $\pi_{C,C'}$ in the range between $\pi_{C,C'} = 0.02$ and $\pi_{C,C'} = 0.32$.

Table 2 displays the criteria pairs that are in SPC. Only the relations between algorithms with different objective functions are shown. There are 24 more criteria

Table 2 Results from
μ-biased ICrA—pairs in SPC

Criteria pairs of ACO algorithms	Value of $\mu_{C,C'}$
$ACOs8 - ACOm8$	0.98
$ACOs5 - ACOm8$	0.96
$ACOs6 - ACOm8$	0.96
$ACOs7 - ACOm8$	0.96
$ACOu3 - ACOm3$	0.95

pairs that are also in SPC. These criteria pairs correspond to the algorithms with the same objective function and close number of ants, for example $ACOu8 - ACOu9$ or $ACOs2 - ACOs3$. As some exceptions, for the criteria pairs $ACOu1 - ACOu2$ and $ACOs4 - ACOs5$, a WPC is monitored and for the $ACOu5 - ACOu6$—WD. It should be noted that these pairs correspond to the case of mono-criteria variant of the problem. These pairs are not shown in the table. There is only one criteria pair corresponding to the multi-criteria case—$ACOm6 - ACOm10$.

The presented criteria pairs show that the SPC is observed for the four cases for $ACOs - ACOm$, and for one—$ACOu - ACOm$. There is not observed strong relation between results using $ACOu$ and $ACOs$. It could be summarized that SPC is only observed between mono-objective algorithms where the number of ants is nominally close, i.e. 5 and 6 ants or 7 and 8 ants. This observation is logical.

In Table 3 again only the results from different objective functions, that are in PC, are displayed. In this case the pairs of multi-criteria algorithms with a nominally close number of ants are appeared, as well as the mono-criteria with a bit larger difference between number of ants (e.g. $ACOs2 - ACOs10$, $ACOs3 - ACOs10$, $ACOu2 - ACOu10$, $ACOu7 - ACOu10$, etc.). The full set of the obtained numerical results could be found at http://intercriteria.net/studies/aco/.

From total 84 cases of criteria pairs between different algorithms 19 pairs are between $ACOs$ and $ACOu$ (mono-criteria) and the rest are between mono and multi-criteria ACO algorithms. That means that the relation between mono and multi-criteria algorithms are weaker than the relations between different mono-criteria algorithms. The last ones are mainly in the SPC.

The presented in Table 3 results show that the algorithms $ACOs - ACOu$ are more similar to each other than the $ACOu - ACOm$ algorithms.

163 criteria pairs are in WPC. 157 from them are between mono and multi-criteria ACO algorithms. Among the results only 6 pairs are between $ACOs - ACOs$ and 4 are between $ACOu - ACOu$. The results show that there os a weaker relation between mono and multi-criteria ACO algorithms. 137 of the criteria pair are between $ACOu - ACOs$ and $ACOu - ACOm$, i.e. $ACOu$ performance is more different in comparison with the performance of the $ACOs$ and $ACOm$ algorithms.

64 criteria pairs are in WD. Among them there are not criteria pairs between $ACOs - ACOs$ algorithms.

In Table 4 the criteria pairs between the same objective function that are in weak positive consonance are presented.

Considering the relations between the same objective function the influence of the different number of ants on the ACO performance is investigated. The results

Table 3 Results from μ-biased ICrA—pairs in PC

Criteria pairs of ACO algorithms	Value of $\mu_{C,C'}$	Criteria pairs of ACO algorithms	Value of $\mu_{C,C'}$
$ACOu6 - ACOm3$	0.93	$ACOs8 - ACOm9$	0.88
$ACOs9 - ACOm8$	0.92	$ACOu2 - ACOm2$	0.88
$ACOu4 - ACOm3$	0.92	$ACOu4 - ACOs1$	0.88
$ACOs8 - ACOm5$	0.92	$ACOu6 - ACOs9$	0.88
$ACOu1 - ACOm1$	0.92	$ACOs9 - ACOm3$	0.87
$ACOu3 - ACOs4$	0.92	$ACOu2 - ACOm4$	0.87
$ACOu4 - ACOm4$	0.92	$ACOu5 - ACOs1$	0.87
$ACOu5 - ACOm3$	0.92	$ACOs9 - ACOm7$	0.87
$ACOu5 - ACOm4$	0.91	$ACOu6 - ACOs10$	0.87
$ACOu3 - ACOm4$	0.91	$ACOs7 - ACOm6$	0.86
$ACOs7 - ACOm5$	0.90	$ACOs8 - ACOm10$	0.86
$ACOs9 - ACOm9$	0.90	$ACOu7 - ACOs10$	0.86
$ACOu10 - ACOm3$	0.90	$ACOs7 - ACOm10$	0.86
$ACOu7 - ACOm3$	0.90	$ACOu10 - ACOs9$	0.86
$ACOu8 - ACOm3$	0.90	$ACOu5 - ACOs10$	0.86
$ACOu9 - ACOm3$	0.90	$ACOu7 - ACOs9$	0.86
$ACOs7 - ACOm7$	0.90	$ACOu8 - ACOs9$	0.86
$ACOu1 - ACOm4$	0.90	$ACOu9 - ACOs9$	0.86
$ACOs7 - ACOm9$	0.89	$ACOu1 - ACOs4$	0.85
$ACOs8 - ACOm7$	0.89	$ACOu2 - ACOm5$	0.85
$ACOu4 - ACOs4$	0.89	$ACOu2 - ACOs10$	0.85
$ACOu2 - ACOm3$	0.89	$ACOu4 - ACOs10$	0.85
$ACOu3 - ACOs1$	0.89	$ACOs9 - ACOm6$	0.85
$ACOu5 - ACOs4$	0.89	$ACOu2 - ACOs4$	0.85
$ACOs8 - ACOm6$	0.88	$ACOu3 - ACOs9$	0.85
$ACOs9 - ACOm5$	0.88	$ACOu6 - ACOs1$	0.85

show that $ACOs$ are less sensitive to the number of ants, compared to the $ACOu$ and $ACOm$. The larger difference of the performance are observed for $ACOu$ and $ACOm$—53% of all criteria pairs. In 14% of the cases the WD is observed for the algorithms $ACOu - ACOu$ and $ACOm - ACOm$ with bigger difference between the number of ants (more than three), **which that** $ACOu$ and $ACOm$ highly influenced of the number of ants.

4.2 Results from Application of v-biased ICrA

All results, based on application of v-biased ICrA, are visualized on Fig. 2 within the specific triangular geometrical interpretation of IFSs.

Table 4 Results from μ-biased ICrA—pairs in WPC

Criteria pairs of ACO algorithms of ACO algorithms	Value of $\mu_{C,C'}$
$ACOs4 - ACOs10$	0.77
$ACOs4 - ACOs5$	0.82
$ACOs4 - ACOs6$	0.82
$ACOs4 - ACOs7$	0.82
$ACOs4 - ACOs8$	0.81
$ACOs4 - ACOs9$	0.84
$ACOu1 - ACOu2$	0.78
$ACOu1 - ACOu3$	0.81
$ACOu1 - ACOu4$	0.82
$ACOu1 - ACOu5$	0.82

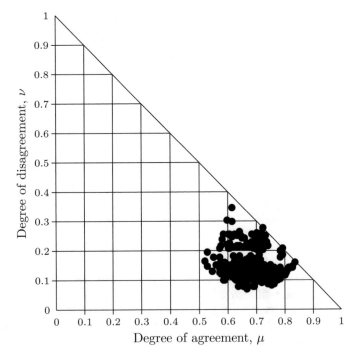

Fig. 2 Presentation of ICrA results in the intuitionistic fuzzy interpretation triangle—ν-biased ICrA

Table 5 Results from v-biased ICrA—pairs in WPC and WD

Criteria pairs of ACO algorithms of ACO algorithms	Value of $\mu_{C,C'}$	Value of $v_{C,C'}$
$ACOs8 - ACOm8$	0.83	0.25
$ACOs5 - ACOm8$	0.71	0.25
$ACOs6 - ACOm8$	0.71	0.25
$ACOs7 - ACOm8$	0.71	0.25
$ACOu3 - ACOm3$	0.81	0.14

The presented results show that in this case the degree of "uncertainty ($v_{C,C'}$) are not zero. The observed $v_{C,C'}$-values are further included in the tables with presented results.

In Table 5 the results for criteria pairs, considered in the case of μ-biased ICrA, are displayed. In comparison with the results in Sect. 4.1, the regarded criteria pairs are not in SPC. There are pairs in WD ($ACOs5 - ACOm8$, $ACOs6 - ACOm8$ and $ACOs7 - ACOm8$) or pairs in WPC ($ACOs8 - ACOm8$ and $ACOu3 - ACOm3$).

The presented results show that in case of v-biased ICrA lower criteria correlations are observed compared to the results based on μ-biased ICrA. The main reason of this is the way of interpretation of equal results in the input index matrix (see http://intercriteria.net/studies/aco/). The equal results are assigned to the degree of "disagreement" and thus the value of $v_{C,C'}$ is increased. In a result, the degree $\mu_{C,C'}$ is decreased.

In Table 6 the results from different objective functions, considered in Sect. 4.1, are presented with the corresponding $v_{C,C'}$-values.

The criteria pairs that are in weak positive consonance are presented in bold. As it can be seen only 18 criteria pairs are in weak positive consonance. The rest of the criteria pairs are in weak dissonance or in dissonance.

In Table 7 the criteria pairs considered in the case of μ-biased ICrA as pairs in WPC are presented. In the case of v-biased ICrA these criteria pairs are in weak dissonance and in dissonance (for criteria pairs $ACOs4 - ACOs10$ and $ACOu1 - ACOu2$).

In Table 8 only the criteria pairs between the same objective function that are in weak positive consonance (v-biased ICrA) are presented. Although, the lower $\mu_{C,C'}$-values the results show that the performance of mono-objective ACO algorithms with sum ($ACOs$) are less sensitive to the number of ants, compared to the mono-objective ACO with multiplication ($ACOu$) and multi-objective ACO ($ACOm$).

4.3 Results from Application of Balanced ICrA

All results, based on application of Balanced ICrA, are visualized on Fig. 3 within the specific triangular geometrical interpretation of IFSs.

Table 6 Results from ν-biased ICrA—pairs in WPC, WD and D

Criteria pairs of ACO algorithms	Value of $\mu_{C,C'}$	Value of $\nu_{C,C'}$	Criteria pairs of ACO algorithms	Value of $\mu_{C,C'}$	Value of $\nu_{C,C'}$
$ACOu6 - ACOm3$	**0.78**	**0.14**	$ACOs8 - ACOm9$	0.67	0.21
$ACOs9 - ACOm8$	0.72	0.20	$ACOu2 - ACOm2$	0.73	0.15
$ACOu4 - ACOm3$	**0.79**	**0.13**	$ACOu4 - ACOs1$	**0.77**	**0.10**
$ACOs8 - ACOm5$	0.71	0.20	$ACOu6 - ACOs9$	0.73	0.15
$ACOu1 - ACOm1$	0.71	0.21	$ACOs9 - ACOm3$	0.74	0.14
$ACOu3 - ACOs4$	**0.80**	**0.12**	$ACOu2 - ACOm4$	**0.76**	**0.11**
$ACOu4 - ACOm4$	**0.78**	**0.13**	$ACOu5 - ACOs1$	**0.77**	**0.10**
$ACOu5 - ACOm3$	0.73	0.11	$ACOs9 - ACOm7$	0.65	0.21
$ACOu5 - ACOm4$	**0.78**	**0.13**	$ACOu6 - ACOs10$	0.69	0.17
$ACOu3 - ACOm4$	**0.79**	**0.13**	$ACOs7 - ACOm6$	0.61	0.25
$ACOs7 - ACOm5$	0.70	0.20	$ACOs8 - ACOm10$	0.61	0.25
$ACOs9 - ACOm9$	0.70	0.21	$ACOu7 - ACOs10$	0.68	0.18
$ACOu10 - ACOm3$	**0.76**	**0.14**	$ACOs7 - ACOm10$	0.60	0.26
$ACOu7 - ACOm3$	**0.76**	**0.14**	$ACOu10 - ACOs9$	0.71	0.15
$ACOu8 - ACOm3$	**0.76**	**0.14**	$ACOu5 - ACOs10$	0.70	0.16
$ACOu9 - ACOm3$	**0.76**	**0.14**	$ACOu7 - ACOs9$	0.71	0.15
$ACOs7 - ACOm7$	0.65	0.25	$ACOu8 - ACOs9$	0.71	0.15
$ACOu1 - ACOm4$	0.71	0.19	$ACOu9 - ACOs9$	0.71	0.15
$ACOs7 - ACOm9$	0.67	0.23	$ACOu1 - ACOs4$	0.70	0.15
$ACOs8 - ACOm7$	0.65	0.24	$ACOu2 - ACOm5$	0.73	0.12
$ACOu4 - ACOs4$	**0.78**	**0.11**	$ACOu2 - ACOs10$	0.70	0.16
$ACOu2 - ACOm3$	**0.77**	**0.11**	$ACOu4 - ACOs10$	0.70	0.16
$ACOu3 - ACOs1$	**0.78**	**0.10**	$ACOs9 - ACOm6$	0.63	0.22
$ACOu5 - ACOs4$	**0.78**	**0.10**	$ACOu2 - ACOs4$	**0.76**	**0.09**
$ACOs8 - ACOm6$	0.63	0.25	$ACOu3 - ACOs9$	0.73	0.12
$ACOs9 - ACOm5$	0.71	0.17	$ACOu6 - ACOs1$	0.75	0.10

In comparison with the results from ν-biased, here the correlation between the criteria pairs preserve the $\mu_{C,C'}$-values. The $\nu_{C,C'}$-values are redistributed between the $\pi_{C,C'}$-values and $\nu_{C,C'}$-values.

Again the obtained in this case results are compared to the results observed in Sect. 4.1. Table 9 displays the $\mu_{C,C'}$ and $\nu_{C,C'}$-values of the criteria pairs considered in case of μ-biased ICrA. In comparison with the results in Sect. 4.1, the regarded criteria pairs are not in SPC, there are in WD ($ACOs5 - ACOm8, ACOs6 - ACOm8$ and $ACOs7 - ACOm8$) or in WPC ($ACOs8 - ACOm8$ and $ACOu3 - ACOm3$).

In Table 10 the results from different objective functions, considered in Sect. 4.1, are displayed with the corresponding $\nu_{C,C'}$-values.

Table 7 Results from v-biased ICrA—pairs in WD and D

Criteria pairs of ACO algorithms	Value of $\mu_{C,C'}$	Value of $v_{C,C'}$
$ACOs4 - ACOs10$	0.66	0.11
$ACOs4 - ACOs5$	0.69	0.13
$ACOs4 - ACOs6$	0.69	0.13
$ACOs4 - ACOs7$	0.69	0.13
$ACOs4 - ACOs8$	0.69	0.12
$ACOs4 - ACOs9$	0.73	0.12
$ACOu1 - ACOu2$	0.66	0.12
$ACOu1 - ACOu3$	0.68	0.13
$ACOu1 - ACOu4$	0.68	0.14
$ACOu1 - ACOu5$	0.68	0.13

Table 8 Results from v-biased ICrA—pairs in WPC

Criteria pairs of ACO algorithms	Value of $\mu_{C,C'}$	Value of $v_{C,C'}$
$ACOu4 - ACOu5$	0.84	0.16
$ACOu3 - ACOu4$	0.83	0.15
$ACOu3 - ACOu5$	0.83	0.14
$ACOu2 - ACOu4$	0.81	0.14
$ACOu2 - ACOu5$	0.81	0.14
$ACOu3 - ACOm3$	0.81	0.14
$ACOu2 - ACOu3$	0.81	0.13

The criteria pairs that are in weak positive consonance and in positive consonance are presented in bold. Compared to the case of v-biased ICrA with 34 criteria pairs in weak dissonance and dissonance, here only 6 pairs are in weak dissonance. In case of μ-biased ICrA all listed in Table 10 criteria pairs are in positive consonance. So, Balanced ICrA shows performance similar to μ-biased ICrA, but still takes account of the repetitive numerical results. Thus, the $\pi_{C,C'}$-values variate from $\pi_{C,C'} = 0.02$ to $\pi_{C,C'} = 0.33$.

In Table 11 the criteria pairs considered in the case of μ-biased ICrA as pairs in WPC are presented. In the case of Balanced ICrA these criteria pairs are in weak dissonance with the exception of three pairs, namely $ACOs4 - ACOs5$, $ACOs4 - ACOs6$ and $ACOs4 - ACOs7$. These criteria pairs are in weak positive consonance. In comparison, in case of μ-biased ICrA, all pairs listed in Table 11 are in weak positive consonance. Applying Balanced ICrA algorithm it is shown that only $ACOs$ algorithm keeps the positive consonance between some of the criteria pairs. Thus, the conclusion that $ACOs$ algorithms are less sensitive to the number of ants, compared to the $ACOu$ and $ACOm$ algorithms, is confirmed under these more stringent conditions.

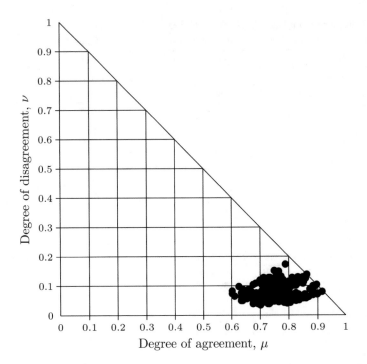

Fig. 3 Presentation of ICrA results in the intuitionistic fuzzy interpretation triangle—Balanced ICrA

Table 9 Results from Balanced ICrA—pairs in WPC and WD

Criteria pairs of ACO algorithms	Value of $\mu_{C,C'}$	Value of $\nu_{C,C'}$
$ACOs8 - ACOm8$	0.86	0.13
$ACOs5 - ACOm8$	0.84	0.12
$ACOs6 - ACOm8$	0.84	0.12
$ACOs7 - ACOm8$	0.84	0.12
$ACOu3 - ACOm3$	0.88	0.07

In the case of Balanced ICrA algorithm there are some criteria pairs between the same objective function that are in positive consonance. The results are presented in Table 12.

In the case of ν-biased ICrA there criteria pairs are in WPC and in the case of μ-biased ICrA they are in PC, even in SPC. If the correlation between given criteria pair (two ACO algorithms) is retained during application of different ICrA approaches, this means that it is really strong dependence of the pair. If the correlation is lost, it means that the resulting numerical results can lead us to an existing relationship.

Table 10 Results from Balanced ICrA—pairs in WPC, PC and WD

Criteria pairs of ACO algorithms	Value of $\mu_{C,C'}$	Value of $\nu_{C,C'}$	Criteria pairs of ACO algorithms	Value of $\mu_{C,C'}$	Value of $\nu_{C,C'}$
$ACOu6 - ACOm3$	0.85	0.07	$ACOs8 - ACOm9$	0.67	0.21
$ACOs9 - ACOm8$	0.76	0.07	$ACOu2 - ACOm2$	0.73	0.15
$ACOu4 - ACOm3$	0.82	0.10	$ACOu4 - ACOs1$	0.83	0.05
$ACOs8 - ACOm5$	0.82	0.10	$ACOu6 - ACOs9$	0.81	0.07
$ACOu1 - ACOm1$	0.81	0.11	$ACOs9 - ACOm3$	0.81	0.07
$ACOu3 - ACOs4$	0.86	0.06	$ACOu2 - ACOm4$	0.82	0.06
$ACOu4 - ACOm4$	0.85	0.07	$ACOu5 - ACOs1$	0.82	0.05
$ACOu5 - ACOm3$	0.85	0.06	$ACOs9 - ACOm7$	0.76	0.11
$ACOu5 - ACOm4$	0.85	0.06	$ACOu6 - ACOs10$	0.78	0.09
$ACOu3 - ACOm4$	0.85	0.06	$ACOs7 - ACOm6$	0.74	0.12
$ACOs7 - ACOm5$	0.80	0.10	$ACOs8 - ACOm10$	0.73	0.13
$ACOs9 - ACOm9$	0.80	0.10	$ACOu7 - ACOs10$	0.77	0.09
$ACOu10 - ACOm3$	0.83	0.07	$ACOs7 - ACOm10$	0.73	0.13
$ACOu7 - ACOm3$	0.83	0.07	$ACOu10 - ACOs9$	0.78	0.07
$ACOu8 - ACOm3$	0.83	0.07	$ACOu5 - ACOs10$	0.78	0.08
$ACOu9 - ACOm3$	0.83	0.07	$ACOu7 - ACOs9$	0.78	0.07
$ACOs7 - ACOm7$	0.77	0.13	$ACOu8 - ACOs9$	0.78	0.07
$ACOu1 - ACOm4$	0.80	0.09	$ACOu9 - ACOs9$	0.78	0.07
$ACOs7 - ACOm9$	0.78	0.11	$ACOu1 - ACOs4$	0.78	0.08
$ACOs8 - ACOm7$	0.77	0.12	$ACOu2 - ACOm5$	0.79	0.06
$ACOu4 - ACOs4$	0.84	0.05	$ACOu2 - ACOs10$	0.77	0.08
$ACOu2 - ACOm3$	0.83	0.06	$ACOu4 - ACOs10$	0.77	0.08
$ACOu3 - ACOs1$	0.84	0.05	$ACOs9 - ACOm6$	0.74	0.11
$ACOu5 - ACOs4$	0.84	0.05	$ACOu2 - ACOs4$	0.81	0.04
$ACOs8 - ACOm6$	0.76	0.13	$ACOu3 - ACOs9$	0.79	0.06
$ACOs9 - ACOm5$	0.80	0.08	$ACOu6 - ACOs1$	0.80	0.05

4.4 Results from Application of Unbiased ICrA

All results, based on application of Unbiased ICrA, are visualized on Fig. 4 within the specific triangular geometrical interpretation of IFSs.

In this case the obtained $\mu_{C,C'}$-values are the same of those obtained in the case of ν-biased ICrA (Sect. 4.2). The difference in the results is in the estimates of $\nu_{C,C'}$-values. All $\nu_{C,C'}$-values are zero, resulting in non zero degree of "uncertainty $\pi_{C,C'}$. The $\pi_{C,C'}$-values variate from $\pi_{C,C'} = 0.16$ to $\pi_{C,C'} = 0.48$. The algorithm of Unbiased ICrA in case of repetitive numerical results in input index matrix increases the value of the degree of "uncertainty $\pi_{C,C'}$. That is the reason of zero estimates for $\nu_{C,C'}$-values.

Table 11 Results from Balanced ICrA—pairs in WD and WPC

Criteria pairs of ACO algorithms	Value of $\mu_{C,C'}$	Value of $\nu_{C,C'}$
$ACOs4 - ACOs10$	0.71	0.06
$ACOs4 - ACOs5$	**0.76**	**0.06**
$ACOs4 - ACOs6$	**0.76**	**0.06**
$ACOs4 - ACOs7$	**0.76**	**0.06**
$ACOs4 - ACOs8$	0.75	0.06
$ACOs4 - ACOs9$	0.79	0.06
$ACOu1 - ACOu2$	0.72	0.06
$ACOu1 - ACOu3$	0.75	0.06
$ACOu1 - ACOu4$	0.75	0.07
$ACOu1 - ACOu5$	0.75	0.07

Table 12 Results from Balanced ICrA—pairs in PC

Criteria pairs of ACO algorithms	Value of $\mu_{C,C'}$	Value of $\nu_{C,C'}$
$ACOu4 - ACOu5$	0.92	0.08
$ACOu3 - ACOu4$	0.90	0.07
$ACOu3 - ACOu5$	0.90	0.07
$ACOu2 - ACOu4$	0.89	0.07
$ACOu2 - ACOu5$	0.88	0.07
$ACOu3 - ACOm3$	0.88	0.07
$ACOu2 - ACOu3$	0.87	0.06

Table 13 Results from μ-biased and Unbiased ICrA

Criteria pairs of ACO algorithms	ν-biased ICrA			Unbiased ICrA		
	$\mu_{C,C'}$	$\nu_{C,C'}$	$\pi_{C,C'}$	$\mu_{C,C'}$	$\nu_{C,C'}$	$\pi_{C,C'}$
$ACOs8 - ACOm8$	0.83	0.25	0.08	0.83	0.00	0.17
$ACOs5 - ACOm8$	0.71	0.25	0.04	0.71	0.00	0.29
$ACOs6 - ACOm8$	0.71	0.25	0.04	0.71	0.00	0.29
$ACOs7 - ACOm8$	0.71	0.25	0.04	0.71	0.00	0.29
$ACOu3 - ACOm3$	0.81	0.14	0.05	0.81	0.00	0.19

The comparison of the here obtained results for different objective functions (criteria pairs in SPC in the case of μ-biased ICrA), with the results in the case of ν-biased ICrA algorithm, is presented in Table 13.

As it be can see the difference in the criteria pairs correlations is only in the interpretation of repetitive numerical results in the input matrix. In the case of μ-biased ICrA algorithm the repetitive numerical results are interpreted as degree of "disagreement", and in the case of Unbiased ICrA algorithm—as degree of "uncertainty".

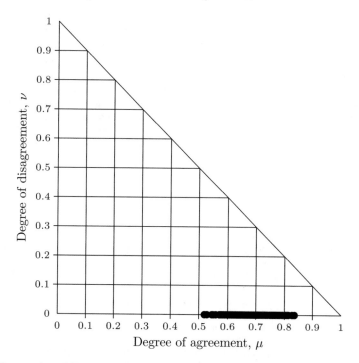

Fig. 4 Presentation of ICrA results in the intuitionistic fuzzy interpretation triangle—Unbiased ICrA

5 Conclusion

The InterCriteria analysis is a powerful tool for studying relations between different objects. We study three variants of ACO algorithm applied on WSN problem. Every variant is tested with various number of ants, between 1 and 10. We search the correlation between variants of ACO and number of ants. WSN problem is a multi-objective problem. When it is converted to mono-objective by summing the two objective functions, the algorithm is less sensitive to the number of used ants. When the problem is solved like multi-objective, we observe bigger difference of algorithm performance according to the number of ants. There is greater similarity between performance of the two mono-objective variants, than between some of mono-objective and multi-objective variants. The InterCriteria analysis is applied in order to examine the relations between regarded hybrid ACO algorithms. The influence of the number of ants is analysed. Four different algorithms of ICrA—μ-biased, Balanced, ν-biased and Unbiased—are applied. The obtained results show that in case of correlation between given hybrid ACO algorithms this dependence is appeared regardless of the ICrA algorithm we use.

In a future we will continue with study of the influence of the ACO parameters values on algorithm performance, taking into account other parameters of ACO, such as evaporation, initial pheromon level, pheromon updating and so on.

Acknowledgements Work presented here is partially supported by the Bulgarian National Scientific Fund under Grants DFNI DN 12/5 "Efficient Stochastic Methods and Algorithms for Large-Scale Problems" and KP-06-N22/1 "Theoretical Research and Applications of InterCriteria Analysis". The development of the proposed hybrid ACO algorithms has been funded by Grant DFNI DN 12/5. The study of influence of number of ants on ACO algorithm behaviour based on ICrA approach has been funded by Grant KP-06-N22/1.

References

1. Atanassov, K.: Generalized index matrices. Comptes rendus de l'Academie Bulgare des Sciences **40**(11), 15–18 (1987)
2. Atanassov, K.: On index matrices, part 1: standard cases. Adv. Stud. Contemp. Math. **20**(2), 291–302 (2010)
3. Atanassov, K.: On index matrices, part 2: intuitionistic fuzzy case. Proc. Jangjeon Math. Soci. **13**(2), 121–126 (2010)
4. Atanassov, K.: Index Matrices: Towards an Augmented Matrix Calculus, Studies in Computational Intelligence, vol. 573 (2014)
5. Atanassov, K.: On Intuitionistic Fuzzy Sets Theory. Springer, Berlin (2012)
6. Atanassov, K., Vassilev, P.: On the intuitionistic fuzzy sets of n-th type. In: Gaweda, A., Kacprzyk, J., Rutkowski, L., Yen, G. (eds.) Advances in Data Analysis with Computational Intelligence Methods. Studies in Computational Intelligence, vol. 738, pp. 265–274. Springer, Cham (2018)
7. Atanassov, K.: Intuitionistic Fuzzy Sets, VII ITKR Session, Sofia, 20–23 June 1983. Reprinted: Int. J. Bioautomation **20**(S1), S1–S6 (2016)
8. Atanassov, K.: Review and New Results on Intuitionistic Fuzzy Sets, Mathematical Foundations of Artificial Intelligence Seminar, Sofia, (1988), Preprint IM-MFAIS-1-88. Reprinted: Int. J. Bioautomation, **20**(S1), S7–S16 (2016)
9. Atanassov, K., Atanassova, V., Gluhchev, G.: InterCriteria analysis: ideas and problems. Notes on Intuitionistic Fuzzy Sets **21**(2), 81–88 (2015)
10. Atanassov, K., Mavrov, D., Atanassova, V.: Intercriteria decision making: a new approach for multicriteria decision making. Based Index Matrices Intuitionistic Fuzzy Sets, Issues in IFSs and GNs **11**, 1–8 (2014)
11. Atanassov, K., Szmidt, E., Kacprzyk, J.: On intuitionistic fuzzy pairs. Notes on Intuitionistic Fuzzy Sets **19**(3), 1–13 (2013)
12. Fidanova, S., Marinov, P., Alba, E.: Ant algorithm for optimal sensor deployment, computational intelligence. In: Madani, K., Correia, A.D., Rosa, A., Filipe, J. (eds.) Studies of Computational Intelligence, vol. 399, pp. 21–29 Springer (2012)
13. Fidanova, S., Marinov, P., Paprzycki, M.: Influence of the number of ants on multi-objective ant colony optimization algorithm for wireless sensor network layout. In: Assessment of the Air Quality in Bulgaria-Short Summary Based on Recent Modelling Results, pp. 232–239 (2014). https://doi.org/10.1007/978-3-662-43880-0_25.
14. Fidanova, S., Marinov, P.: Influence of the number of ants on mono-objective ant colony optimization algorithm for wireless sensor network layout. In: Proceedings of BGSIAM'12, pp. 59–66, Sofia, Bulgaria (2012)
15. Fidanova, S., Shindarov, M., Marinov, P.: Wireless sensor positioning using ACO algorithm. In: Sgurev, V., Yager, R., Kacprzyk, J., Atanassov, K. (eds.) Recent Contributions in Intelligent Systems, Studies in Computational Intelligence, Vol. 657, pp. 33–44. Springer, Cham (2017)

16. Fidanova, S., Roeva, O., Paprzycki, M., Gepner, P.: InterCriteria analysis of ACO start strategies. In: Proceedings of the 2016 Federated Conference on Computer Science and Information Systems, pp. 547–550 (2016)
17. Hernandez, H., Blum, C.: Minimum energy broadcasting in wireless sensor networks: an ant colony optimization approach for a realistic antenna model. J. Appl. Soft Comput. 11(8), 5684–5694 (2011)
18. Ikonomov, N., Vassilev, P., Roeva, O.: ICrAData—software for intercriteria analysis. Int J Bioautomation 22(1), 1–10 (2018)
19. D.B. Jourdan, Wireless Sensor Network Planning with Application to UWB Localization in GPS-denied Environments, Massachusets Institute of Technology, Ph.D. thesis (2000)
20. Konstantinidis, A., Yang, K., Zhang, Q., Zainalipour-Yazti, D.: A multiobjective evolutionary algorithm for the deployment and power assignment problem in wireless sensor networks. J. Comput. Netw. 54(6), 960–976 (2010)
21. Krawczak, M., Bureva, V., Sotirova, E., Szmidt, E.: Application of the InterCriteria decision making method to universities ranking. Adv. Intell. Syst. Comput. 401, 365–372 (2016)
22. Molina, G., Alba, E., Talbi, E.-G.: Optimal sensor network layout using multi-objective metaheuristics. Univers. Comput. Sci. 14(15), pp. 2549–2565 (2008)
23. Marinov, E., Vassilev, P., Atanassov, K.: On separability of intuitionistic fuzzy sets, In: Novel Developments in Uncertainty Representation and Processing, Advances in Intelligent Systems and Computing, Vol. 401, pp. 111–123. Springer, Cham (2106)
24. Ribagin, S., Shannon, A., Atanassov, K.: Intuitionistic fuzzy evaluations of the elbow joint range of motion. Adv. Intell. Syst. Comput. 401, 225–230 (2016)
25. Roeva, O., Vassilev, P., Ikonomov, N., Angelova, M., Su, J., Pencheva, T.: On Different Algorithms for InterCriteria Relations Calculation, Studies in Computational Intelligence, Vol. 757, pp. 143–160. Springer, Cham (2019)
26. Pottie, G.J., Kaiser, W.J.: Embedding the internet: wireless integrated network sensors. Commun. ACM 43(5), 51–58 (2000)
27. Todinova, S., Mavrov, D., Krumova, S., Marinov, P., Atanassova, V., Atanassov, K., Taneva, S.G.: Blood plasma thermograms dataset analysis by means of InterCriteria and correlation analyses for the case of colorectal cancer. Int. J. Bioautomation 20(1), 115–124 (2016)
28. Traneva, V., Atanassova V., Tranev S.: Index matrices as a decision-making tool for job appointment. In: Nikolov, G. et al. (Ed.) Springer Nature Switzerland AG, pp. 1–9. NMA 2018, LNCS 11189 (2019)
29. Traneva, V., Tranev, S., Atanassova, V.: An Intuitionistic fuzzy approach to the hungarian algorithm. In: Nikolov, G. et al. (Ed.) Springer Nature Switzerland AG, pp. 1-9, NMA 2018, LNCS 11189, (2019)
30. Vassilev, P.: A Note on New Distances between Intuitionistic Fuzzy Sets, Notes on Intuitionistic Fuzzy Sets, Vol. 21, No. 5, 11–15 (2015)
31. Vassilev, P., Todorova, L., Andonov, V.: An auxiliary technique for InterCriteria Analysis via a three dimensional index matrix, Notes on Intuitionistic Fuzzy Sets, vol. 21, No. 2, 71–76 (2015)
32. Vassilev, P., Ribagin, S.: A note on intuitionistic fuzzy modal-like operators generated by power mean. In: Kacprzyk J., Szmidt E., Zadrony S., Atanassov K., Krawczak M. (eds.) Advances in Fuzzy Logic and Technology 2017 (2018). EUSFLAT 2017, IWIFSGN, : Advances in Intelligent Systems and Computing, vol. 643, pp. 470–475. Springer, Cham (2017)
33. Wolf, S., Mezz, P.: Evolutionary local search for the minimum energy broadcast problem. In: Cotta, C., van Hemezl, J. (eds.) VOCOP 2008. Lecture Notes in Computer Sciences, vol. 4972, pp. 61–72. Springer, Germany (2008)
34. Xu, Y., Heidemann, J., Estrin, D.: Geography informed energy conservation for Ad Hoc routing. In: Proceedings of the 7th ACM/IEEE Annual International Conference on Mobile Computing and Networking, Italy, pp. 70–84 (2001)

Geometric Versus Arithmetic Difference and H^1-Seminorm Versus L^2-Norm in Function Approximation

Stefan M. Filipov, Ivan D. Gospodinov, Atanas V. Atanassov and Jordanka A. Angelova

Abstract In solving function approximation problems, sometimes, it is necessary to ensure non-negativity of the result and approximately uniform relative deviation from the target. In these cases, it is appropriate to use the geometric difference to define distance. On the other hand, sometimes, it is more important that the result looks like the target, rather than being close to it in Euclidean sense. Then, to define distance, it is more appropriate to use the H^1-seminorm rather than the L^2-norm. This paper considers four possible metrics (distance functions): (i) L^2-norm of the arithmetic difference, (ii) L^2-norm of the geometric difference, (iii) H^1-seminorm of the arithmetic difference, and (iv) H^1-seminorm of the geometric difference. Using the four different metrics, we solve the following function approximation problem: given a target function and a set of linear equality constraints, find the function that satisfies the constraints and minimizes the distance, in the considered metric, to the target function. To solve the problem, we convert it to a finite dimensional constrained optimization problem by discretizing it. Then, the method of Lagrange multipliers is applied. The obtained systems for case (i) and (iii) are linear and are solved exactly. The systems for case (ii) and (iv), however, are nonlinear. To solve them, we propose a self-consistent iterative procedure which is a combination of fixed-point iteration and Newton method. Particular examples are solved and discussed.

S. M. Filipov (✉) · I. D. Gospodinov · A. V. Atanassov
Department of Computer Science, University of Chemical Technology and Metallurgy,
blvd. Kl. Ohridski 8, 1756 Sofia, Bulgaria
e-mail: filipovstefan@yahoo.com

I. D. Gospodinov
e-mail: gospodinov@uctm.edu

A. V. Atanassov
e-mail: naso@uctm.edu

J. A. Angelova
Department of Mathematics, University of Chemical Technology and Metallurgy, blvd. Kl.
Ohridski 8, 1756 Sofia, Bulgaria
e-mail: jordanka_aa@yahoo.com

© Springer Nature Switzerland AG 2020
S. Fidanova (ed.), *Recent Advances in Computational Optimization*,
Studies in Computational Intelligence 838,
https://doi.org/10.1007/978-3-030-22723-4_7

Keywords Constrained optimization · Lagrange multipliers · Distance · Dissimilarity · Closeness · Similarity

1 Introduction

Broadly speaking, the function approximation problem asks us to choose, from a set of well-defined functions, the function that most closely matches, in a task specific way, a given target function [1–6]. In this work, the functions from which we must choose the sought function are defined by a number of equality constraints. If a function satisfies (meets) the constraints, then it is a member of the set of allowed functions. The constrains that we consider are only linear: linear combination of functional values at certain points, Dirichlet boundary conditions, and linear integral conditions.

In order to measure how closely a function from the set of allowed functions matches the given target function we need to introduce a distance (metric). The chosen metric depends on our needs. If we want the magnitude of the functional difference between the sought and the given target function to be as small as possible and as uniformly distributed across the domain as possible, then we normally choose as a metric the L^2-norm (Euclidean norm) of the arithmetic difference between the two functions. This leads to the well-known least squares method [7–9].

Sometimes, however, given a positive target function, the sought function is required to be positive too. In addition, to the extent to which the constraints allow, the magnitude of the arithmetic difference between the sought and the given function, across the domain, is required to be proportional to the given function itself. In this case we could use as a metric the L^2-norm of the geometric difference [10–12], i.e. the logarithm of the ratio, between the sought and the given function. This metric ensures that the sought function is positive and that, roughly speaking, the magnitude of the relative difference (percentage difference) between the sought and the given function is as small and as uniformly distributed across the domain as possible [12].

Sometimes, the sought function is required to be as similar to the given target function as possible, i.e. the two functions are required to look alike rather than being close to each other in Euclidean sense. In other words, to the extent to which the constrains allow, the sought function should be equal to the given target function plus some constant. In this case we could use as a metric the H^1-seminorm of the arithmetic difference [13]. Various applications of this metric can be found in [14–18].

Finally, given a positive target function, the sought function is required to be positive and as similar in a relative sense as possible. In other words, to the extent to which the constrains allow, the sought function should be equal to the given target function multiplied by some constant. In this case we could use as a metric the H^1-seminorm of the geometric difference [19].

In this paper we solve the formulated above function approximation problem using the discussed four different metrics. In order to solve the problem, we convert it to a finite dimensional constrained optimization problem by discretizing it. Then we apply the method of Lagrange multipliers [20]. For two of the cases, those involving the arithmetic difference, the obtained systems are linear and are solved directly. For the other two cases, those involving the geometric difference, the obtained systems are nonlinear. To solve these nonlinear systems, we propose a self-consistent iterative procedure which is a combination of fixed-point iteration and Newton method.

This paper is structured as follows. In Sect. 2 the arithmetic difference and the geometric difference between the two functions are considered. The absolute distance is defined as the L^2-norm of the arithmetic difference. The relative distance is defined as the L^2-norm of the geometric difference. Then, we consider the H^1-seminorm of the arithmetic difference. The closer to zero it is, the more absolutely similar the two functions are. Finally, we consider the H^1-seminorm of the geometric difference. The closer to zero it is, the more relatively similar the two functions are. In Sect. 3 we formulate the function approximation problem. In Sect. 4 the problem is discretized and converted to a finite dimensional constrained optimization problem. In Sect. 5 the problem is solved by the method of Lagrange multipliers. In Sect. 6 a self-consistent iterative procedure for the solution of the nonlinear systems is proposed. Numerical results are presented in Sect. 7.

2 Arithmetic Difference, Geometric Difference, L^2-Norm, and H^1-Seminorm

Let $u : [a, b] \rightarrow \mathbb{R}$ be some given target function and let $u^* : [a, b] \rightarrow \mathbb{R}$ be another function defined on the same interval. The arithmetic difference $\delta : [a, b] \rightarrow \mathbb{R}$ between u^* and u at point $t \in [a, b]$ is

$$\delta(t) = u^*(t) - u(t).$$ (1)

If $\delta(t)$ is a constant $C \in \mathbb{R}$ throughout the interval, then

$$u^*(t) = u(t) + C,$$ (2)

and the two functions are equally separated at each point t. Example is shown in Fig. 1.

If $\delta(t) = 0$ in the whole interval, then $u^*(t) \equiv u(t)$, i.e. the two functions are identical. Thus, the absolute distance between u^* and u can be defined as:

(i) L^2-norm of the arithmetic difference:

$$\|u^* - u\| = \sqrt{\int_a^b (u^*(t) - u(t))^2 \, dt}.$$ (3)

Fig. 1 The functions u^* and u are equally separated from each other at each point t in an absolute sense

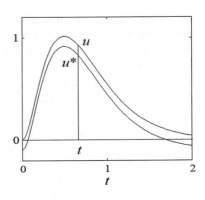

Fig. 2 The functions u^* and u are equally separated from each other at each point t in a relative sense

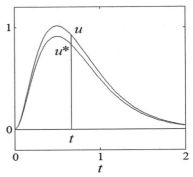

Now, let u^* and u be two positive functions. The geometric difference $\delta_L : [a, b] \to \mathbb{R}$ between u^* and u at point $t \in [a, b]$ (see [10–12]) is

$$\delta_L(t) = \ln \frac{u^*(t)}{u(t)} = \ln u^*(t) - \ln u(t). \tag{4}$$

If $\delta_L(t)$ is a constant $C \in \mathbb{R}$ throughout the interval, then the two functions are equally separated at each point t in a relative sense,

$$u^*(t) = e^C u(t). \tag{5}$$

Example is shown in Fig. 2.

If $\delta_L = 0$ in the whole interval, then $u^*(t) \equiv u(t)$, i.e. the two functions are identical. Expanding $\ln u^*$ in the right-hand side of (4) in a Taylor series around u gives $\delta_L = \delta_R + O(\delta_R^2)$, where $\delta_R = \delta/u$ is the relative difference. Hence, for small δ_R the geometric difference δ_L is a measure of the relative difference between the functions u^* and u. Thus, the relative distance between u^* and u can be defined as:

Fig. 3 The functions u^* and u are completely similar in an absolute sense: $u^* = u + const$

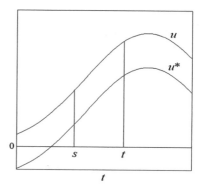

Fig. 4 The functions u^* and u are completely similar in a relative sense: $u^* = const \cdot u$

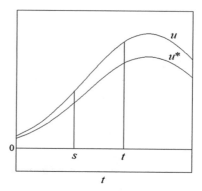

(*ii*) L^2-*norm of the geometric difference*:

$$\left\| \ln u^* - \ln u \right\| = \sqrt{\int_a^b (\ln u^*(t) - \ln u(t))^2 \, dt}. \tag{6}$$

Let us now consider the case when the two functions u^* and u are not required to be close to each other in Euclidean sense but rather to be similar, i.e. look alike, in absolute (Fig. 3) or relative (Fig. 4) sense.

Let $t, s \in [a, b]$ be two arbitrary points from the interval. From $u^* = u + const$ (see Fig. 3) it can be inferred that

$$u^*(t) - u(t) = u^*(s) - u(s). \tag{7}$$

Fixing the point s and taking the derivative of (7) with respect to t we get:

$$\frac{du^*(t)}{dt} - \frac{du(t)}{dt} = 0. \tag{8}$$

Therefore, equality of the derivatives throughout the interval can be used as a condition for complete absolute similarity of the two functions. Thus, as a measure of absolute dissimilarity between the two functions we can take:

(*iii*) H^1-*seminorm of the arithmetic difference*:

$$\left|u^* - u\right|_{H^1} = \sqrt{\int_a^b \left(\frac{du^*(t)}{dt} - \frac{du(t)}{dt}\right)^2 dt}. \tag{9}$$

The closer to zero (9) is, the more absolutely similar the two functions are.

From $u^* = const \cdot u$ (see Fig. 4), it can be inferred that

$$\frac{u^*(t)}{u(t)} = \frac{u^*(s)}{u(s)}. \tag{10}$$

Taking the logarithm of (10), fixing the point s, and then taking the derivative with respect to t, we get:

$$\frac{d \ln u^*(t)}{dt} - \frac{d \ln u(t)}{dt} = 0. \tag{11}$$

Therefore, equality (11) can be used as a condition for complete relative similarity of the two functions. Thus, as a measure of relative dissimilarity between the two functions, we can take:

(*iv*) H^1-*seminorm of the geometric difference*:

$$\left|\ln u^* - \ln u\right|_{H^1} = \sqrt{\int_a^b \left(\frac{d \ln u^*(t)}{dt} - \frac{d \ln u(t)}{dt}\right)^2 dt}. \tag{12}$$

The closer to zero (12) is, the more relatively similar the two functions are.

3   Statement of the Function Approximation Problem

Let $u : [a, b] \to \mathbb{R}$ be some given bounded target function. Let d be one of the four distances defined in Table 1 and let D be the set of all functions $\bar{u} : [a, b] \to \mathbb{R}$ for which $d(\bar{u}, u)$ is defined. Let $G \subset D$ be the set of functions $\bar{u}$ that satisfy the given constraints, namely: a certain number of linear integral constraints of the form

$$\int_a^b f(t)\bar{u}(t)dt = 1, \tag{13}$$

where $f : [a, b] \to \mathbb{R}$ is a weight function, and possibly one or two boundary conditions of the form

Table 1 Four possible metrics. The smaller the distance/dissimilarity is, the closer/more similar the two functions are

	Arithmetic difference	Geometric difference
L^2	$d(u^*, u) = \|u^* - u\|$ (i) absolute distance	$d(u^*, u) = \|\ln u^* - \ln u\|$ (ii) relative distance
H^1	$d(u^*, u) = |u^* - u|_{H^1}$ (iii) absolute dissimilarity	$d(u^*, u) = |\ln u^* - \ln u|_{H^1}$ (iv) relative dissimilarity

$$\mu \bar{u}(a) + v \bar{u}(b) = \gamma, \tag{14}$$

where $\mu, v, \gamma \in \mathbb{R}$ and $\mu^2 + v^2 > 0$.

We seek the function $u^* \in G$ which is at a minimal distance from the target, i.e.:

$$d(u^*, u) = \min_{\bar{u} \in G} d(\bar{u}, u). \tag{15}$$

There is no restriction to the total number of constraints as long as they are consistent with each other and the formulated problem is well posed.

4 Discretizing the Problem

The interval $[a, b]$ is partitioned by N mesh-points t_i, $i = 1, 2, \ldots, N$ into $N - 1$ subintervals of equal size. This partitioning defines a uniform mesh on the interval. The values of u^* and u at the meth-points t_i are denoted by u_i^* and u_i and are collected into two $N \times 1$ column-vectors $\mathbf{u}^*$ and $\mathbf{u}$, sometimes called mesh-functions. The sought values u_i^*, $i = 1, 2, \ldots, N$ should satisfy M linear equations called constraints. The number of constraints M should be less than N. The constraints can be written in a matrix form:

$$\mathbf{A} \cdot \mathbf{u}^* = \mathbf{c}, \tag{16}$$

where $\mathbf{c}$ is an $M \times 1$ column-vector and $\mathbf{A}$ is an $M \times N$ matrix. Note that any integral constraint (13) can be incorporated in (16) by replacing the integral with a corresponding sum. For this purpose, we need to use some formula for numerical integration.

Now, we discretize the distances (metrics) given in Table 1. Instead of the norms / semi-norms we use their squares since it is more convenient. For the first derivatives we use their finite difference approximations. The integrals are replaced by sums and any constant factors are removed since they do not affect the minimization. The obtained in this way objective functions are given in Table 2.

Given a target column-vector (mesh-function) $\mathbf{u}$, we have to find the column-vector (mesh-function) $\mathbf{u}^*$ that satisfies the linear constrains (16) and minimizes

Table 2 The objective functions I corresponding to the four different metrics given in Table 1

	Arithmetic difference	Geometric difference
L^2	$I = \sum_{i=1}^{N} \left(u_i^* - u_i\right)^2$ (*i*) discretized square of *absolute distance*	$I = \sum_{i=1}^{N} \left(\ln u_i^* - \ln u_i\right)^2$ (*ii*) discretized square of *relative distance*
H^1	$I = \sum_{i=1}^{N-1} \left(\left(u_{i+1}^* - u_i^*\right) - \left(u_{i+1} - u_i\right)\right)^2$ (*iii*) discretized square of *absolute dissimilarity*	$I = \sum_{i=1}^{N-1} \left(\left(\ln u_{i+1}^* - \ln u_i^*\right) - \left(\ln u_{i+1}\right.\right.$ $\left.\left. - \ln u_i\right)\right)^2$ (*iv*) discretized square of *relative dissimilarity*

the objective function I. For objective function we can choose any of the objective functions given in Table 2. Formulated in this way, the problem is a finite dimensional constrained optimization problem for the unknown u_i^*, $i = 1, 2, \ldots, N$.

5 Method of Lagrange Multipliers

To solve the problem, we apply the method of Lagrange Multipliers [20]. The constraints are given by the matrix equation (16). It is a system of M linear algebraic equations. First, the left-hand side of (16) is transferred to the right. Then, the j-th equation from the system is multiplied by the Lagrange multiplier λ_j, $j = 1, 2, \ldots, M$, summed over j, and the result is added to the objective function I to obtain the Lagrangian

$$J = I + \sum_{j=1}^{M} \lambda_j \left(c_j - \sum_{i=1}^{N} A_{ji} u_i^*\right). \tag{17}$$

Then, the derivatives of J with respect to the unknowns u_k^*, $k = 1, 2, \ldots, N$ should be equated to zero

$$\frac{\partial J}{\partial u_k^*} = 0, \; k = 1, 2, \ldots, N. \tag{18}$$

For objective function I, we can choose any of the objective functions given in Table 2.

First, let I be the discretized square of the absolute distance, i.e. case (i) in Table 2. Equations (18) give:

$$2\left(u_k^* - u_k\right) - \sum_{j=1}^{M} \lambda_j A_{jk}, \; k = 1, 2, \ldots, N. \tag{19}$$

The system (19) can be written in the matrix equation form

Table 3 The sought vector (mesh-function) $\mathbf{u}^*$ as function of the Lagrange multipliers λ for the four different metrics

	Arithmetic difference	Geometric difference
L^2	$\mathbf{u}^* = \mathbf{u} + \frac{1}{2}\mathbf{A}^T \cdot \lambda$ (i) *absolutely closest* mesh-function	$\mathbf{u}^* = \mathbf{u}\exp\left(\frac{1}{2}\mathbf{u}^*\left(\mathbf{A}^T \cdot \lambda\right)\right)$ (ii) *relatively closest* mesh-function
H^1	$\mathbf{u}^* = \mathbf{u} -$ $\left(\bar{\mathbf{L}} + \bar{\mathbf{A}}\right)^{-1}\left(\frac{1}{2}\mathbf{A}^T \cdot \lambda + \bar{\mathbf{A}} \cdot \mathbf{u} - \bar{\mathbf{c}}\right)$ (iii) *absolutely most similar* mesh-function	$\mathbf{u}^* = \mathbf{u}\exp\left(-\frac{1}{2}\left(\bar{\mathbf{L}} + \mathbf{U}\right)^{-1} \cdot \mathbf{u}^*\left(\mathbf{A}^T \cdot \lambda\right)\right)$ (iv) *relatively most similar* mesh-function

$$\mathbf{u}^* = \mathbf{u} + \frac{1}{2}\mathbf{A}^T.\lambda. \tag{20}$$

The result (20) is given in Table 3 case (i). The derivations for case (ii), (iii), and (iv) are presented in [12, 13, 19], respectively. The results are given in Table 3.

In Table 3 $\lambda = [\lambda_1, \lambda_2, \ldots, \lambda_M]^T$ is the vector with the Lagrange multipliers, $\bar{\mathbf{A}}$ is the matrix $\mathbf{A}$ augmented to $N \times N$ matrix by adding $N - M$ rows of zeros, $\bar{\mathbf{c}}$ is the vector $\mathbf{c}$ augmented to $N \times 1$ vector by adding $N - M$ zeros [13], $\bar{\mathbf{L}}$ and $\mathbf{U}$ are the following $N \times N$ matrices [19]:

$$\bar{\mathbf{L}} = \begin{pmatrix} -1 & 1 & 0 & 0 & \ldots & 0 & 0 & 0 \\ 1 & -2 & 1 & 0 & \ldots & 0 & 0 & 0 \\ 0 & 1 & -2 & 1 & \ldots & 0 & 0 & 0 \\ \vdots & \vdots & \vdots & \vdots & \ldots & \vdots & \vdots & \vdots \\ 0 & 0 & 0 & 0 & \ldots & 1 & -2 & 1 \\ 0 & 0 & 0 & 0 & \ldots & 0 & 1 & -1 \end{pmatrix} \quad \mathbf{U} = \begin{pmatrix} u_1 & u_2 & \ldots & u_N \\ 0 & 0 & \ldots & 0 \\ \vdots & \vdots & \ddots & \vdots \\ 0 & 0 & \ldots & 0 \end{pmatrix}. \tag{21}$$

To write conveniently the matrix equations for the geometric difference, i.e. case (ii) and (iv) in Table 3, we use the following definition:

Definition 1 Let $\mathbf{x}$ and $\mathbf{y}$ be two $N \times 1$ column-vectors: $\mathbf{x} = [x_1, x_2, \ldots, x_N]^T$, $\mathbf{y} = [y_1, y_2, \ldots, y_N]^T$. We define the no-sign product between $\mathbf{x}$ and $\mathbf{y}$ as

$$\mathbf{xy} = [x_1 y_1, x_2 y_2, \ldots, x_N y_N]^T$$

and the exponent of a vector as

$$\exp(\mathbf{x}) = \left[e^{x_1}, e^{x_2}, \ldots, e^{x_N}\right]^T.$$

Note that, although the matrix $\bar{\mathbf{L}}$ is singular, the matrices $\bar{\mathbf{L}} + \bar{\mathbf{A}}$ and $\bar{\mathbf{L}} + \mathbf{U}$ are (in general) regular [13, 19]. The formula (iv) in Table 3 is derived for cases when one of the constraints is area preserving, i.e. $\sum_{i=1}^{N} u_i^* = \sum_{i=1}^{N} u_i$. This, however, is not

Table 4 The lagrange
multipliers λ for cases (i) and
(iii) in Table 3

	Arithmetic difference
L^2	$\lambda = 2\left(\mathbf{A} \cdot \mathbf{A}^T\right)^{-1} \cdot (\mathbf{c} - \mathbf{A} \cdot \mathbf{u})$ (i)
H^1	$\lambda = 2\left(\mathbf{A} \cdot (\bar{\mathbf{L}} + \bar{\mathbf{A}})^{-1} \cdot \mathbf{A}^T\right)^{-1} \cdot$
	$\left(\mathbf{A} \cdot \mathbf{u} - \mathbf{c} - \mathbf{A} \cdot (\bar{\mathbf{L}} + \bar{\mathbf{A}})^{-1} \cdot (\bar{\mathbf{A}} \cdot \mathbf{u} - \bar{\mathbf{c}})\right)$ (iii)

restrictive as long as one of the constraints for u^* is of the form $\sum_{i=1}^{N} u_i^* = s$. In this case, any given target u_{target} should first be multiplied by a number to convert it to a target u that satisfies $\sum_{i=1}^{N} u_i = s$. Then, this new target u can be used in formula (iv) Table 3 (for details see [19]). Note that the new target u is completely relatively similar to the original target u_{target} because it is obtained through multiplication by a number.

The matrix equation (20) and the matrix equation for the constrains (16) are $N + M$ equations for the $N + M$ unknowns u_i^*, $i = 1, 2, \ldots, N$ and λ_j, $j = 1, 2, \ldots, M$. To find the vector of the Lagrange multipliers $\lambda = [\lambda_1, \lambda_2, \ldots, \lambda_M]^T$ we plug $\mathbf{u}^*$ from (20) into (16) and solve for λ:

$$\lambda = 2\left(\mathbf{A}.\mathbf{A}^T\right)^{-1}. (\mathbf{c} - \mathbf{A}.\mathbf{u}). \tag{22}$$

The result (22) is given in Table 4 case (i). The result for case (iii) is also given in Table 4. The derivations are presented in [13]. Substituting λ from (22) into (20), we finally get

$$\mathbf{u}^* = \mathbf{u} + \mathbf{A}^T \cdot \left(\mathbf{A} \cdot \mathbf{A}^T\right)^{-1} \cdot (\mathbf{c} - \mathbf{A} \cdot \mathbf{u}). \tag{23}$$

Similarly, we can obtain $\mathbf{u}^*$ for case (iii) by substituting λ (iii) from Table 4 into the equation for $\mathbf{u}^*$ (iii) Table 3. We note that, although the formulae look complicated, they are very easy to implement in, for example, MATLAB.

For cases (ii) and (iv), after substituting the corresponding $\mathbf{u}^*$ from Table 3 into the equation for the constraints (16), we get nonlinear equation which cannot be solved directly. In the next paragraph we propose a self-consistent iterative procedure for determining $\mathbf{u}^*$ and λ simultaneously.

6 Self-consistent Iterative Procedure

In this paragraph we propose an iterative procedure for finding the relatively closest (ii) and the relatively most similar (iv) mesh-function $\mathbf{u}^*$ (Table 3). We consider case (ii), but case (iv) is analogous and the results for it are given in the corresponding tables.

Table 5 Equations obtained by substituting $\mathbf{u}^*$ (ii) and (iv) from Table 3 into the equation for the constraints (16). In the table $\mathbf{H} = (\bar{\mathbf{L}} + \mathbf{U})^{-1}$

	Geometric difference
L^2	$\mathbf{A} \cdot \mathbf{u} \exp\left(\frac{1}{2}\mathbf{u}^* \left(\mathbf{A}^T \cdot \boldsymbol{\lambda}\right)\right) = \mathbf{c}$ (ii)
H^1	$\mathbf{A} \cdot \mathbf{u} \exp\left(-\frac{1}{2}\mathbf{H}.\mathbf{u}^* \left(\mathbf{A}^T.\boldsymbol{\lambda}\right)\right) = \mathbf{c}$ (iv)

Table 6 Components of the vector $\mathbf{e}(\boldsymbol{\lambda})$ for case (ii) and case (iv). In the table $\mathbf{H} = (\bar{\mathbf{L}} + \mathbf{U})^{-1}$

	Geometric difference
L^2	$e_j = c_j - \sum_{i=1}^{N} A_{ji} u_i \exp\left(\frac{1}{2} u_i^* \left(\mathbf{A}^T.\boldsymbol{\lambda}\right)_i\right), \quad j = 1, 2, \ldots, M$ (ii)
H^1	$e_j = c_j - \sum_{i=1}^{N} A_{ji} u_i \exp\left(-\frac{1}{2} (\mathbf{H}.\mathbf{u}^*)_i \left(\mathbf{A}^T.\boldsymbol{\lambda}\right)_i\right), \quad j = 1, 2, \ldots, M$ (iv)

First, $\mathbf{u}^*$ (ii) Table 3 is substituted in the equation for the constraints (16):

$$\mathbf{A} \cdot \mathbf{u} \exp\left(\frac{1}{2}\mathbf{u}^* \left(\mathbf{A}^T.\boldsymbol{\lambda}\right)\right) = \mathbf{c}. \tag{24}$$

The analogous equation for case (iv) is given in Table 5.

The unknown $\mathbf{u}^*$ and $\boldsymbol{\lambda}$ should satisfy equation (24) and equation (ii) Table 3. We propose the following procedure for solving both equations simultaneously. First, we choose an initial guess for $\boldsymbol{\lambda}$ and solve equation (ii) Table 3 iteratively, using fixed-point iteration, to obtain the first approximation for $\mathbf{u}^*$. It is convenient to use fixed-point iteration since $\mathbf{u}^*$ is expressed as a function of $\mathbf{u}^*$. The obtained $\mathbf{u}^*$ is substituted into (24) and the equation is solved iteratively, using the Newton method, to obtain the first approximation of $\boldsymbol{\lambda}$. This $\boldsymbol{\lambda}$ is substituted into equation (ii) Table 3 and the equation is solved again to obtain the second approximation of $\mathbf{u}^*$, and so on until $\mathbf{u}^*$ and $\boldsymbol{\lambda}$ stop changing. To implement the Newton method, we proceed as follows. First, we transfer the left-hand side of (24) to the right and introduce the $M \times 1$ column-vector $\mathbf{e}(\boldsymbol{\lambda})$ with components given in (ii) Table 6. Solving (24) is equivalent to solving $\mathbf{e}(\boldsymbol{\lambda}) = 0$. This can be done using the Newton iteration:

$$\boldsymbol{\lambda}_{new} = \boldsymbol{\lambda} - \left(\frac{\partial \mathbf{e}}{\partial \boldsymbol{\lambda}}\right)^{-1} .\mathbf{e}(\boldsymbol{\lambda}). \tag{25}$$

The components of the Jacobian are given in (ii) Table 7. The procedure for case (iv) is the same but $\mathbf{e}(\boldsymbol{\lambda})$ and the Jacobian are given in (iv) Table 6 and Table 7, respectively.

Table 7 Components of the Jacobian $\partial \mathbf{e}(\lambda)/\partial \lambda$ for case (ii) and (iv). In the table $\mathbf{H} = \left(\bar{\mathbf{L}} + \mathbf{U}\right)^{-1}$

	Geometric difference
L^2	$\dfrac{\partial e_j}{\partial \lambda_m} = -\dfrac{1}{2} \sum\limits_{i=1}^{N} A_{ji} u_i \, \exp\left(\dfrac{1}{2} u_i^* \left(\mathbf{A}^T.\lambda\right)_i\right) u_i^* A_{mi}, \quad j, m = 1, 2, \ldots, M \ (ii)$
H^1	$\dfrac{\partial e_j}{\partial \lambda_m} = \dfrac{1}{2} \sum\limits_{i=1}^{N} A_{ji} u_i \, \exp\left(-\dfrac{1}{2} (\mathbf{H}.\mathbf{u}^*)_i \left(\mathbf{A}^T.\lambda\right)_i\right) (\mathbf{H}.\mathbf{u}^*)_i A_{mi}, \quad j, m =$ $1, 2, \ldots, M \ (iv)$

7 Numerical Examples and Discussion

In this section three numerical examples are shown. The first two demonstrate the benefits of using relative as opposed to absolute closeness, i.e. case (ii) vs case (i), when the target represents intrinsically nonnegative quantity. In such cases the result (the sought function) must be nonnegative. In addition, it is preferable that the result be, to the extent to which the constraints allow, proportional to the magnitude of the target across the domain. The third example compares absolute with relative similarity, i.e. case (iii) and case (iv). These two metrics are good when the two functions must look alike rather than being close to each other in functional values (in Euclidean distance). Again, when the target in nonnegative and the result is required to be nonnegative, relative similarity is a better choice than absolute.

Example 1 L^2-norm of arithmetic difference vs L^2-norm of geometric difference.

The target-function u is given on the interval $[0, 3]$ as the following mesh-function (vector):

$$u_i = 30 t_i^2 \exp\left(-4t_i\right), \quad t_i = \frac{3\,(i-1)}{40}, \quad i = 1, 2, \ldots, 41. \tag{26}$$

The sought function u^* must satisfy the following two linear integral constraints:

$$\sum_{i=1}^{N} u_i^* = 10.62, \quad \sum_{i=1}^{N} \exp\left(t_i\right) u_i^* = 25.63. \tag{27}$$

We find the absolutely closest to u function u^* (i) Table 3, denoted by u_a, and the relatively closest to u function u^* (ii) Table 3, denoted u^*, by minimizing, subject to constrains (27), the objective functions (i) Table 2 and (ii) Table 2, respectively. The results are shown in Fig. 5. As can be seen u^* is nonnegative everywhere and its overall behavior captures the behavior of the target very well. The function u_a is negative at zero and in $2 < t < 2.4$. Its behavior differs qualitatively from that of the target function. For example, towards the end of the interval, instead of decreasing, u_a increases. On the other hand u_a is closer to u in functional values.

Fig. 5 Solution to
Example 1. The target u
(26), the absolutely closets to
it function u_a, and the
relatively closest to it
function u^* that satisfy the
constraints (27) are shown

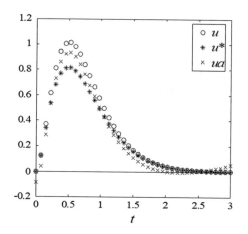

Example 2 L^2-norm of arithmetic difference vs L^2-norm of geometric difference.

The target-function u is given on the interval $[0, 2\pi]$ as the following mesh-function (vector):

$$u_i = \cos^2 (t_i) \exp\left(-\frac{t_i}{2}\right), \quad t_i = \frac{\pi (i-1)}{20}, \quad i = 1, 2, \dots, 41. \quad (28)$$

The sought function u^* must satisfy the following two linear integral constraints:

$$\sum_{i=1}^{N} u_i^* = 0.8 \sum_{i=1}^{N} u_i, \quad \sum_{i=1}^{N} (t_i^2 - 1) u_i^* = 1.2 \sum_{i=1}^{N} (t_i^2 - 1) u_i. \quad (29)$$

As in Example 1 we find the absolutely closest to u function u^* (i) Table 3, denoted by u_a, and the relatively closest to u function u^* (ii) Table 3, denoted u^*, by minimizing, subject to constrains (29), the objective functions (i) Table 2 and (ii) Table 2, respectively (Fig. 6).

Again, the function u_a is closer to the target function u in absolute values but it is negative at some places and its overall behavior does not agree very well with that of the target function. The function u^*, however, is nonnegative everywhere and behaves similarly to the target.

Example 3 H^1-seminorm of arithmetic difference vs H^1-seminorm of geometric difference.

When the constrains require that at least one of the values of the target, e.g. a boundary value, must be changed considerably, then looking for closeness in Euclidean sense is not a good option. Then, however, we can seek similarity between the sought function and the given target function. Again, we have two options. We can choose either absolute or relative similarity, depending on the requirements.

The original target function is given as mesh-function on the interval $[-1, 1]$. It is denoted by u_o:

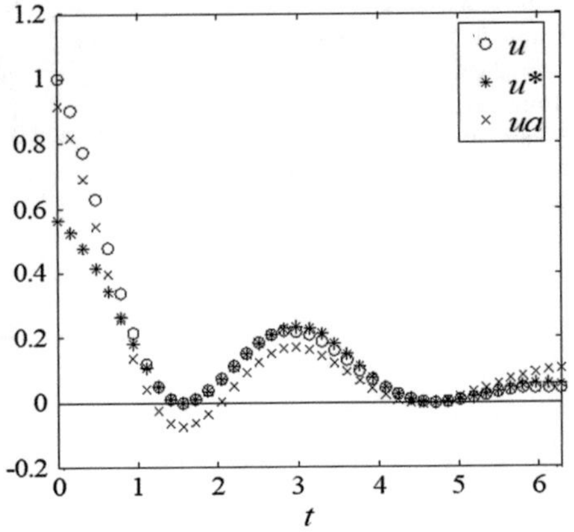

Fig. 6 Solution to
Example 2. The target u
(28), the absolutely closets to
it function u_a, and the
relatively closest to it
function u^* that satisfy the
constraints (29) are shown

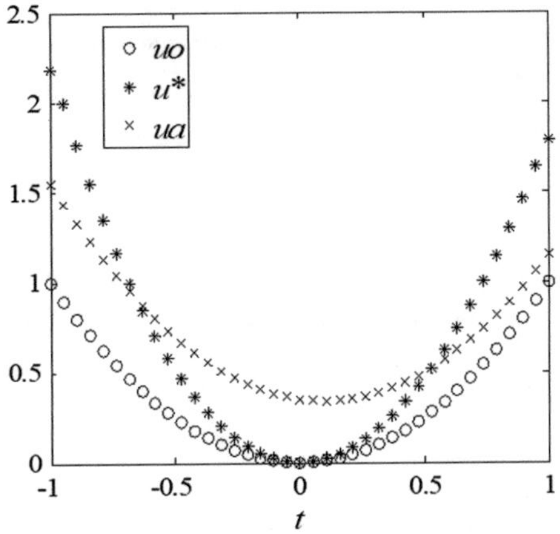

Fig. 7 Solution to
Example 3. The original
target u_o (30), the absolutely
most similar to it function
u_a, and the relatively most
similar to it function u^* that
satisfy the constraints (31),
(32) are shown

$$u_{oi} = t_i^2, \quad t_i = -1 + \frac{i-1}{19}, \quad i = 1, 2, \ldots, 39. \tag{30}$$

The constraints that the sought function u^* must satisfy are

$$\sum_{i=1}^{N} u_i^* = 2 \sum_{i=1}^{N} u_{oi}, \tag{31}$$

$$u_1^* - u_N^* = 0.4. \tag{32}$$

Constraint (31) is linear integral constraint, while constraint (32) is a difference constraint. It can be viewed as nonlocal boundary condition.

We find the most absolutely similar to the target u function u^* (iii) Table 3, denoted by u_a, and the most relatively similar u^* (iv) Table 3, denoted by u^*, by minimizing, subject to constraints (31), (32), the objective functions (iii) and (iv) in Table 2, respectively. For case (iv) we first convert the original target u_o to the target $u = 2u_o$ (see Sect. 5). To solve the nonlinear system for case (iv) we use the self-consistent iterative procedure described in Sect. 6. The results are shown in Fig. 7..

As can be seen from Fig. 7, the absolutely most similar function u_a looks like $u + const$, while the relatively most similar function u^* looks like $const \cdot u$. This is in complete agreement with what is expected, given the definitions of complete similarity in an absolute and relative sense given in Sect. 2 Fig. 3 and Fig. 4, respectively.

8 Conclusion

This work considered four possible metrics that can be used to measure the closeness/similarity between two function: (i) The L^2-norm of the arithmetic difference. The smaller it is, the closer the two functions are in an absolute sense; (ii) The L^2-norm of the geometric difference. The smaller it is, the closer the two functions are in a relative sense; (iii) The H^1-seminorm of the arithmetic difference. The smaller it is, the more similar the two functions are in an absolute sense; (iv) The H^1-seminorm of the geometric difference. The smaller it is, the more similar the two function are in a relative sense. The considered function approximation problem was solved, for all four cases, by first converting it to a finite dimensional constrained optimization problem and then using the Lagrange multipliers method. For cases (ii) and (iv), when geometric difference is concerned, the resulting systems are nonlinear. They were solved by applying a self-consistent iterative procedure. The obtained results indicate that metrics (ii) and (iv) guarantee nonnegativity of the results for any nonnegative original target function. In addition, to the extent to which the constraints allow, the result is deviated from the target by (approximately) the same percentage throughout the entire interval.

References

1. Achiezer, N.I.: Theory of Approximation (translated by C. J. Hyman). Ungar, New York (1956)
2. Timan, A.F.: Theory of Approximation of Functions of a Real Variable. Pergamon (Translated from Russian) (1963)
3. Korneichuk, N.P., Ligun, A.A., Doronin, V.G.: Approximation with Constraints. Kiev (In Russian) (1982)
4. Milovanović, G.V., Wrigge, S.: Least Squares Approximation with Constraints. Math. Comput. (AMS) **46**(174), 551–565 (1986)
5. Lorentz, G.G.: Approximation of Functions. AMS (2005)
6. Pedregal, P.: Optimization and Approximation. Springer International Publishing (2017)
7. Gauss, C.F.: Theory of the Combination of Observations Least Subject to Errors. (Translated from original 1820 manuscript by G. W. Stewart). Society for Industrial and Applied Mathematics (1995)

8. Lawson, C.L., Hanson, R.J.: Solving Least Squares Problems. SIAM (1995)
9. Bjorck, A.: Numerical Methods for Least Squares Problems. Society for Industrial and Applied Mathematics (1996)
10. Graff, C.: Expressing relative differences (in percent) by the difference of natural logarithms. J. Math. Psych. **60**, 82–85 (2014)
11. Graff, C.: Why estimating relative differences by Ln(A/B) in percentage and why naming it geometric difference. In: Conference: ICPS, Amsterdam, Netherlands (2015). https://hal. archives-ouvertes.fr/hal-01480972/document
12. Filipov, S.M., Atanasov, A.V., Gospodinov, I.D.: Solving function approximation problems using the L^2-norm of the log ratio as a metric. In: Nikolov, G., et al. (eds.) Numerical Methods and Applications. NMA 2018. Lecture Notes in Computer Science, vol. 11189, pp. 115-124. Springer Cham (2019). https://doi.org/10.1007/978-3-030-10692-8_13
13. Filipov, S.M., Atanasov, A., Gospodinov, I.D.: Constrained functional similarity by minimizing the H^1 seminorm and applications to engineering problems. J. Sci. Eng. Educ. **1**(1), 61-67 (2016). http://dl.uctm.edu/see/node/jsee2016-1/12-Filipov_61-67.pdf
14. Filipov, S.M., Atanasov, A., Gospodinov, I.D.: Constrained similarity of 2D trajectories by minimizing the H^1 semi-norm of the trajectory difference. Optimization and Control. arXiv: 1608.08541 [math.OC] https://arxiv.org/abs/1608.08541
15. Filipov, S.M., Nikolov, V., Gospodinov, I.D.: Constrained similarity of surfaces by minimizing the L^2 norm of the gradient of the surface difference. Industry 4.0 (STUME) **2**, 78-80 (2016). https://stumejournals.com/journals/i4/2016/2/78
16. Gospodinov, I.D., Krumov, K., Filipov, S.M.: Laplacian preserving transformation of surfaces and application to boundary value problems for Laplaces and Poissons equations. Math. Model. (STUME) **1**(1), 14–17 (2017). https://stumejournals.com/journals/mm/2017/1/14
17. Filipov, S.M., Gospodinov, I.D., Farag, I.: Shooting-projection method for two-point boundary value problems. Appl. Math. Lett. **72**, 10–15 (2017). https://doi.org/10.1016/j.aml.2017.04.002
18. Filipov, S.M., Gospodinov, I.D., Angelova, J.: Solving two-point boundary value problems for integro-differential equations using the simple shooting-projection method. In: Dimov, I., Fidanova, S., Lirkov, I. (eds.) NMA 2014. LNCS, vol. 8962, pp. 169–177. Springer, Cham (2015). https://doi.org/10.1007/978-3-319-15585-2_19
19. Filipov, S.M., Atanasov, A.V., Gospodinov, I.D.: Constrained relative functional similarity by minimizing the H^1 semi-norm of the logarithmic difference. In: International Conference Automatics 2016. Proceedings of Technical University of Sofia, vol. 66, no. 2, pp. 349–358 (2016). http://proceedings.tu-sofia.bg/volumes/Proceedings_volume_66_book_2_2016.pdf
20. Bertsekas, D.P.: Constrained Optimization and Lagrange Multiplier Methods. Massachusetts Institute of Technology, Athena Scientific, Belmont (1996)

Ellipsoidal Estimates of Reachable Sets for Nonlinear Control Systems with Bilinear Uncertainty

Tatiana F. Filippova

Abstract The estimation problems for control systems with unknown but bounded uncertainties are studied. The uncertainty description is of a set-membership type when it is assumed that only some upper bounds for unknown items are given. The state estimation approaches based on special structure of nonlinearity and uncertainty that are simultaneously present in the control system are developed. The studies are motivated by numerous modeling problems for dynamical systems with uncertainty and nonlinearity in different fields such as physical engineering problems, economical modeling and finances, ecological problems, demographical issues etc. This investigation continues previous researches and a more complicated case is considered here, when the dynamical equations of control system contain two types of nonlinearities, one of which is of quadratic type and another one contains uncertain matrix parameters. Such models may arise in applications related, in particular, to satellite control problems with nonlinearity and disturbances in the model description. New results presented here consist in deriving the dynamical equations for the ellipsoidal estimates of reachable sets of the control system under study. Related numerical algorithms and simulation results are also given.

1 Introduction

The estimation problems for uncertain dynamical systems are considered here for the case when a probabilistic description of noise and errors is not available, but only bounds on them are known [1–6]. Mathematical models of such systems appear in studies of dynamical models under uncertainty in many applications such as physics, cybernetics, biology, economics and other areas [7–13]. Therefore, the development

T. Filippova (✉)
Krasovskii Institute of Mathematics and Mechanics, Russian Academy of Sciences,
16 S. Kovalevskaya str., Yekaterinburg 620990 , Russian Federation
e-mail: ftf@imm.uran.ru

© Springer Nature Switzerland AG 2020
S. Fidanova (ed.), *Recent Advances in Computational Optimization*,
Studies in Computational Intelligence 838,
https://doi.org/10.1007/978-3-030-22723-4_8

of analytical methods and numerical schemes for the analysis of control systems with nonlinearity and uncertainty is very important for both theory and related applications.

Here the modified state estimation techniques which use the special structure of nonlinearity of a dynamical system and also take into account state constraints are proposed. We assume that the system nonlinearity is generated by the combination of two types of functions in related differential equations, one of which is bilinear and the other one is quadratic. The additional state constraints (of ellipsoidal type) are also imposed.

The paper further develops previous basic results of [14–22] and also continues some recent studies [23–27] in this field, namely we give here the new versions of the upper estimates of reachable sets of the control system with uncertainty, nonlinearity and state constraints. It is worth mentioning here also the similar results based on the the modified state estimation approach connected with the Hamilton-Jacobi-Bellman (HJB) equation techniques which also allow to construct the external ellipsoidal estimates of reachable sets [28, 29].

2 Problem Formulation

2.1 Basic Notations

Let $\mathbb{R}^n$ denote the n–dimensional Euclidean space and $x'y$ is the usual inner product of $x, y \in \mathbb{R}^n$ with the prime as a transpose and with $\|x\| = (x'x)^{1/2}$. We use the symbol comp $\mathbb{R}^n$ for the variety of all compact subsets $A \subset \mathbb{R}^n$ and the symbol conv $\mathbb{R}^n$ for the variety of all compact convex subsets $A \subset \mathbb{R}^n$.

Let us denote the set of all closed convex subsets $A \subseteq \mathbb{R}^n$ by the symbol clconv $\mathbb{R}^n$. Let $\mathbb{R}^{n \times m}$ stands for the set of all real $n \times m$-matrices, diag$\{v\}$ denotes a diagonal matrix with the elements of vector v on the main diagonal. Denote by $I \in \mathbb{R}^{n \times n}$ the identity matrix and by Tr (A) the trace of $n \times n$-matrix A (the sum of its diagonal elements).

We denote also by $B(a, r) = \{x \in \mathbb{R}^n : \|x - a\| \le r\}$ the ball in $\mathbb{R}^n$ with a center $a \in \mathbb{R}^n$ and a radius $r > 0$ and denote by

$$E(a, Q) = \{x \in \mathbb{R}^n : (Q^{-1}(x - a), (x - a)) \le 1\}$$

the *ellipsoid* in $\mathbb{R}^n$ with a center $a \in \mathbb{R}^n$ and with a symmetric positive definite $n \times n$-matrix Q.

2.2 Main Problem Description

Consider the following nonlinear control system under uncertainty conditions

$$\dot{x} = A(t)x + f(x)d + u(t), \quad x_0 \in \mathcal{X}_0, \quad t \in [t_0, T], \tag{1}$$

where $x, d \in \mathbb{R}^n$, $\|x\| \leq K$ $(K > 0)$, $f(x)$ is the nonlinear function, which is quadratic in x, that is $f(x) = x'Bx$, with a given symmetric and positive definite $n \times n$-matrix B.

Control function $u(t)$ is assumed to be Lebesgue measurable on $[t_0, T]$ and it satisfies the constraint $u(t) \in \mathcal{U}$ for a.e. $t \in [t_0, T]$ where $\mathcal{U}$ is a given set belonging to comp $\mathbb{R}^n$. We will assume further that $\mathcal{U} = E(\hat{a}, \hat{Q})$.

We assume that the $n \times n$-matrix function $A(t)$ in (1) is a sum of two specific matrix functions,

$$A(t) = A^0 + A^1(t), \tag{2}$$

where A^0 is a given constant $n \times n$-matrix and a measurable (in t) $n \times n$-matrix $A^1(t)$ is unknown but bounded, $A^1(t) \in \mathcal{A}^1$ $(t \in [t_0, T])$.

Therefore taking into account the above constraint we may rewrite it in the following form

$$A(t) \in \mathcal{A} = A^0 + \mathcal{A}^1, \tag{3}$$

where we assume that

$$\mathcal{A}^1 = \big\{ A = \{a_{ij}\} \in R^{n \times n} : a_{ij} = 0 \text{ for } i \neq j, \text{ and}$$
$$a_{ii} = a_i, \quad i = 1, \ldots, n, \quad a = (a_1, \ldots, a_n), \quad a'Da \leq 1 \big\}, \tag{4}$$

with $D \in \mathbb{R}^{n \times n}$ being a symmetric and positive definite matrix.

We assume further that $\mathcal{X}_0$ in (1) is an ellipsoid, $\mathcal{X}_0 = E(a_0, Q_0)$, with a symmetric and positive definite matrix $Q_0 \in \mathbb{R}^{n \times n}$ and with a center $a_0 \in \mathbb{R}^n$.

One of the important issues in the theory of control under uncertainty is how to specify the set of all solutions $x(t)$ to (1) that satisfy the additional state constraints (the "viability" constraint [1, 2])

$$x(t) \in Y(t), \quad t_0 \leq t \leq T \tag{5}$$

where $Y(t) \in \text{conv } \mathbb{R}^n$ $(t \in [t_0, T])$.

In particular, the viability constraint (5) may be induced by the so-called measurement equation

$$y(t) = G(t)x + w(t),$$

where $y(t)$ is a p-vector function corresponding to measurement results which are obtained with unknown but bounded "noises" $w(t)$

$$w \in Q^*(t), \quad Q^*(t) \in \text{comp } \mathbb{R}^p,$$

here $Q^*(t)$ is a given set-valued function, $G(t)$ is a given $p \times n$–matrix function, (some earlier problem settings and previous results in this field may be found in [2]).

Here we consider the case when the state constraint of type (5) is defined by the ellipsoid,

$$x(t) \in Y = E(\tilde{a}, \tilde{Q}), \quad t_0 \le t \le T, \tag{6}$$

with the center $\tilde{a} \in \mathbb{R}^n$ and the positive definite $n \times n$–matrix $\tilde{Q}$. We will assume further that there exists at least one solution $x^*(t)$ of (1) that satisfies the condition (6).

Let the absolutely continuous function $x(t) = x\big(t; u(\cdot), A(\cdot), x_0\big)$ be a solution to dynamical system (1)–(6) with initial state $x_0 \in \mathcal{X}_0$, with admissible control $u(\cdot)$ and with a matrix $A(\cdot)$ satisfying (2)–(4). The reachable set $\mathcal{X}(t)$ at time t $(t_0 < t \le T)$ of system (1)–(5) is defined as the set

$$\mathcal{X}(t) = \Big\{ x \in \mathbb{R}^n : \ \exists x_0 \in \mathcal{X}_0, \ \exists u(\cdot) \in \mathcal{U}, \ \exists A(\cdot) \in \mathcal{A}, \\ x = x(t) = x\big(t; u(\cdot), A(\cdot), x_0\big), \ x(s) \in Y, \ t_0 \le s \le t \Big\} \tag{7}$$

It should be noted that the problem of exact construction of reachable sets of control systems is very difficult even for linear systems [1, 3]. So instead of exact problem solution the different estimation approaches were developed in many researches [1–5]. The main problem studied here is to further develop the ellipsoidal techniques and to find the effective external ellipsoidal estimates (with respect to inclusion of sets) for reachable sets $\mathcal{X}(t)$ $(t_0 < t \le T)$. It is very important also to study the dynamics in time t of such upper ellipsoidal estimates because they may be used as the indicators of the system dynamics. We investigate here a more complicated case than in [23], since we assume now that we have an additional state constraint on the trajectories of the control system, which significantly complicates the problem analysis.

3   Main Results

Earlier some approaches were proposed to obtain differential equations describing dynamics of external (and in some cases internal) ellipsoidal estimates for reachable sets of control system under uncertainty, e.g., in [12, 24] the authors studied estimation problems for systems with uncertain matrices in dynamical equations, but additional nonlinear terms in dynamics were not considered there.

Differential equations of ellipsoidal estimates for reachable sets of a nonlinear dynamical control system were derived in [16] for the case when system state velocities contain quadratic forms but in that case the uncertainty in matrix coefficients was not assumed.

Later, in [18], differential equations for external ellipsoidal estimates of reachable sets of a control system with nonlinearity and with uncertain matrix were derived under assumption that all elements $\{a_{ij}\}$ of the matrix A were bounded in modulus.

Here we investigate the case different from the above mentioned results when we assume the quadratic-type constraints on unknown matrix $A(t)$ included in the system dynamics and assume also the presence of additional constraints on the system states. In this case the related analysis of the dynamical properties of proposed ellipsoidal estimates is more complicated and was not carried out before.

3.1 Auxiliary Estimate for Systems with State Constraints

We use here the idea of [2] for the so-called elimination of state constraints in the construction of reachable sets of the system (1)–(6) (see also related results in [25]). Consider the following auxiliary differential inclusion with $n \times n$–matrix parameter L,

$$\dot{z} \in (A^0 - L + \mathcal{A}^1)z + f(z) \cdot d + E(\hat{a}, \hat{Q}) + L \cdot E(\tilde{a}, \tilde{Q}),$$

$$t_0 \le t \le T, \quad z_0 \in X_0 = E(a_0, Q_0). \tag{8}$$

Denote by $Z(t; t_0, X_0, L)$ ($t \in [t_0, t_1]$) the trajectory tube to (8) for a fixed matrix parameter L. Let $\mathcal{L}$ denotes the class of all $n \times n$-matrix L.

We will need the following result which may be used as a basis for the constraints elimination procedure.

Lemma 1 *The following estimate is true*

$$X(t) \subseteq \bigcap_{L \in \mathcal{L}} Z(t; t_0, X_0, L), \quad t_0 \le t \le T. \tag{9}$$

Proof The proof follows the main lines and ideas presented in [2]. It was proven even more strong result in [2] that more precise upper estimates similar to (9) may be obtained if we will use a wider class of matrices L (for example, if we consider all matrices depending on time, $L = L(t)$), here for simplicity we consider only constant matrices L.

Using this approach and taking into account results of [16, 18, 19, 22, 23] we can find the upper ellipsoidal estimates for reachable sets $Z(t) = Z(t; t_0, X_0, L)$ of the nonlinear system (8).

3.2 Main Theorem

The following result describes the dynamics of the external ellipsoidal estimates of the reachable set $X(t) = X(t; t_0, X_0)$ ($t_0 \le t \le T$) of the system (1)–(6).

Denote the maximal eigenvalue of the matrix $B^{1/2}Q_0B^{1/2}$ as k^2, therefore k^2 is the smallest positive number for which the inclusion

$$X_0 = E(a_0, Q_0) \subseteq E(a_0, k^2 B^{-1})$$

is satisfied (see also related constructions in [22]).

Theorem 1 *The inclusion*

$$X(t; t_0, X_0) \subseteq E(a_L^+(t), r_L^+(t)B^{-1}) \tag{10}$$

is true for any $t \in [t_0, T]$ and for any $L \in \mathcal{L}$ where functions $a_L^+(t)$, $r_L^+(t)$ are the solutions of the following system of ordinary differential equations

$$\dot{a}_L^+(t) = (A^0 - L)a_L^+(t) + ((a_L^+(t))'Ba_L^+(t) +$$

$$r_L^+(t))d + \hat{a} + L\tilde{a}, \quad t_0 \leq t \leq T,$$

$$\dot{r}_L^+(t) = \max_{\|l\|=1} \left\{ l'\left(2r_L^+(t)B^{1/2}(A^0 - L + \right. \right.$$

$$2d(a_L^+(t))'B)B^{-1/2} + \tag{11}$$

$$\left. q^{-1}(r_L^+(t))B^{1/2}\hat{Q}_L^* B^{1/2})\right)l \right\} + q(r_L^+(t))r_L^+(t),$$

$$q(r) = ((nr)^{-1}\mathrm{Tr}(B\hat{Q}^*))^{1/2},$$

where the positive definite matrix $\hat{Q}^*$ is such that

$$\mathcal{A}^1 a_0 + E(0, \hat{Q}) + k_0 D^{1/2}B^{1/2}B(0, 1) + LE(0, \tilde{Q}) \subseteq E(0, \hat{Q}^*) \tag{12}$$

and the initial states for $a_L^+(t)$ and $r_L^+(t)$ are

$$a_L^+(t_0) = a_0, \quad r_L^+(t_0) = k^2.$$

Proof The above estimates are derived following the scheme of the proof of Theorem 2 in [19] and using the result of Lemma 1 with the necessary corrections done according to new class of constraints on unknown parameters and functions included in the system description.

Corollary 1 *The following estimate is true for any $t \in [t_0, T]$*

$$X(t; t_0, X_0) \subseteq \bigcap_{L \in \mathcal{L}} E(a_L^+(t), r_L^+(t)B^{-1}).$$

Proof The inclusion follows directly from Theorem 1.

Remark 1 The numerical scheme and the related algorithm for constructing upper estimates of reachable sets of the system under consideration may be also formulated similar to algorithms described in [16–18].

4 Numerical Simulations

4.1 Particular Case: Nonlinear Uncertain Systems without State Constraints

Consider some examples which show that in nonlinear case the reachable sets of the control system of the studied type (with simultaneously presenting nonlinearity and uncertainty) may lose the convexity property with increasing time $t > t_0$. Nevertheless the related external estimates calculated on the basis of above ideas and results are ellipsoids (and therefore convex) and these ellipsoids contain the true reachable sets of the studied nonlinear system. Note also that these ellipsoidal estimates in some directions are tight that is, they cannot be further reduced, otherwise they will stop evaluating from above (with respect to the inclusion) the real reachable sets.

Example 1. Consider the following control system

$$\begin{cases} \dot{x}_1 = a_1(t)x_1 + x_1^2 + x_2^2 + u_1, \\ \dot{x}_2 = a_2(t)x_2 + u_2, \end{cases} \tag{13}$$

Here we take $x_0 \in X_0 = B(0, 1)$, $0 \leq t \leq 0.4$ and $U = B(0, 0.1)$. The function $a(t) = (a_1(t), a_2(t))$ is unknown, with the constraint $||a(t)|| \leq 0.5$ ($t_0 \leq t \leq T$).

The projections of estimating ellipsoids $E^+(t) = E(a^+(t), Q^+(t))$ (indicated in red lines) and also the reachable sets $X(t)$ (indicated in blue lines) onto the plane of state coordinates are shown at Fig. 1 for time moments $t = 0.08; 0.2; 0.32; 0.38$.

4.2 General Case: Nonlinear Uncertain Systems with State Constraints

Consider now the following nonlinear control system with state constraints.
Example 2.

$$\begin{cases} \dot{x}_1 = a_1 x_1 + x_1^2 + x_2^2 + u_1, \\ \dot{x}_2 = a_2 x_2 + u_2, \\ \quad x_0 \in X_0, \quad t_0 \leq t \leq T. \end{cases} \tag{14}$$

Here $t_0 = 0$, $T = 0.4$. The uncertain initial state x_0 belongs to the ball $X_0 = B(0, 1)$, uncertain parameters $\{a_1, a_2\}$ satisfy the constraint $a_1^2 + a_2^2 \leq 1$, admissible control functions belong to the set $U = B(0, 0.1)$.

Fig. 1 Reachable sets $X(t)$
and its upper estimates
$E(a^+(t), Q^+(t))$

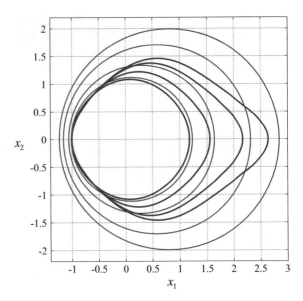

The state constraint is defined by the ellipsoid $E(\tilde{a}, \tilde{Q})$ where $\tilde{a} = 0$ and $\tilde{Q} =$ diag $\{0.64, 4\}$.

The reachable set $X(t)$ with the estimating ellipsoids $E(a_{L_i}(t), Q_{L_i}(t))$ found by Theorem 1 are shown in Fig. 2 for different choices of matrices L_i ($i = 1, 2, 3$, $t = 0.1$).

Fig. 2 Ellipsoidal estimates
for reachable set X(t)
(2d-picture in the plane of
state variables)

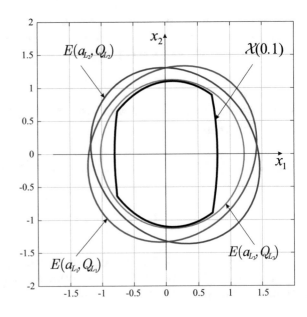

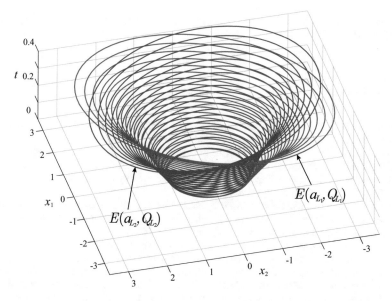

Fig. 3 Ellipsoidal tubes containing reachable set X(t) (3d-picture in the plane of state variables and time)

Two estimating ellipsoidal tubes described in Theorem 1 are shown in Fig. 3.

5   Conclusions

The paper deals with the problems of state estimation for uncertain dynamical control systems under the assumption that the initial system state and some parameters in dynamical equations are unknown but bounded with given constraints.

The focus of this research was done on deriving differential equations which describe the dynamics of the ellipsoidal estimates of reachable sets of the systems under study.

Basing on the results of ellipsoidal calculus developed earlier we present the new state estimation results which use the special bilinear–quadratic structure of nonlinearity and uncertainty of the control system and allow to construct the external ellipsoidal estimates of reachable sets. Examples and numerical results related to procedures of set-valued approximations of trajectory tubes and reachable sets are also given.

The results obtained here may be used in further theoretical and applied researches in optimal control and state estimation for dynamical systems with more complicated classes of uncertainty of set-membership and other types.

Acknowledgements The research was supported by the Russian Science Foundation under RSF Project No.16-11-10146.

References

1. Kurzhanski, A.B. , Varaiya, P.L.: Dynamics and control of trajectory tubes: theory and computation. Syst. Control, Found. Appl. **85** (2014). Basel, Birkhäuser
2. Kurzhanski, A.B., Filippova, T.F.: On the theory of trajectory tubes a mathematical formalism for uncertain dynamics, viability and control. In: Kurzhanski, A.B. (ed.) Advances in Nonlinear Dynamics and Control: A Report from Russia. Progress in Systems and Control Theory, vol. 17, pp. 122–188. Birkhäuser, Boston (1993)
3. Chernousko, F.L.: State Estimation for Dynamic Systems. CRC Press, Boca Raton (1994)
4. Polyak, B.T., Nazin, S.A., Durieu, C., Walter, E.: Ellipsoidal parameter or state estimation under model uncertainty. Automatica **40**, 1171–1179 (2004)
5. Schweppe, F.C.: Uncertain Dynamical Systems. Prentice-Hall, Englewood Cliffs, NJ (1973)
6. Krastanov, M.I., Veliov, V.M.: High-order approximations to nonholonomic affine control systems. In: Large-Scale Scientific Computing 7th International Conference, LSSC 2009, Sozopol, Bulgaria, June 4–8, 2009. Lecture Notes in Computer Science, vol. 5910, 294–301 (2010)
7. Apreutesei, N.C.: An optimal control problem for a prey-predator system with a general functional response. Appl. Math. Lett. **22**(7), 1062–1065 (2009)
8. August E., Lu J., Koeppl H. (2012) Trajectory enclosures for nonlinear systems with uncertain initial conditions and parameters. In: Proceedings of 2012 American Control Conference, Fairmont Queen Elizabeth, Montréal, Canada, Jun. 1488–1493
9. Boscain, U., Chambrion, T., Sigalotti, M.: On some open questions in bilinear quantum control. European Control Conference (ECC), July 2013, pp. 2080–2085. Zurich, Switzerland (2013)
10. Ceccarelli N., Di Marco M., Garulli A., Giannitrapani A. (2004) A set theoretic approach to path planning for mobile robots. In: Proc. 43rd IEEE Conference on Decision and Control, Atlantis, Bahamas, Dec., pp. 147-152
11. Mazurenko, S.S.: A differential equation for the gauge function of the star-shaped attainability set of a differential inclusion. Doklady Math. Mosc. **86**(1), 476–479 (2012)
12. Sinyakov, V.V.: Method for computing exterior and interior approximations to the reachability sets of bilinear differential systems. Differ. Equ. **51**(8), 1097–1111 (2015)
13. Malyshev, V.V., Tychinskii, YuD: Construction of attainability sets and optimization of maneuvers of an artificial Earth satellite with thrusters in a strong gravitational field, Proc. of RAS. Theory Control Syst. **4**, 124–132 (2005)
14. Filippova, T.F.: Set-valued solutions to impulsive differential inclusions. Math. Comput. Model. Dyn. Syst. **11**(2), 149–158 (2005)
15. Filippova T.F. (2010) Construction of set-valued estimates of reachable sets for some nonlinear dynamical systems with impulsive control. In: Proceedings of the Steklov Institute of Mathematics, vol. S.2, 95–102
16. Filippova, T.F.: Differential equations of ellipsoidal state estimates in nonlinear control problems under uncertainty. In: Discrete and Continuous Dynamical Systems, Supplement,: Dynamical Systems, Differential Equations and Applications, vol. 1, pp. 410–419. American Institute of Mathematical Sciences, Springfield (2011)
17. Filippova, T.F. (2015). State estimation for uncertain systems with arbitrary quadratic nonlinearity. In: Proceedings of PHYSCON 2015, Istanbul, Turkey, pp. 1–6
18. Filippova, T.F.: Estimates of reachable sets of impulsive control problems with special nonlinearity. AIP Conf. Proc. **1773**(100004), 1–8 (2016). https://doi.org/10.1063/1.4964998
19. Filippova, T.F.: Differential equations of ellipsoidal state estimates for bilinear-quadratic control systems under uncertainty. J. Chaotic Modeling Simul. (CMSIM) **1**, 85–93 (2017)

20. Filippova, T.F., Berezina, E.V.: On state estimation approaches for uncertain dynamical systems with quadratic nonlinearity: theory and computer simulations. In: Lirkov, I., Margenov, S., Wasniewski, J. (eds.) Large-Scale Scientific Computing. LSSC 2007. Lecture Notes in Computer Science, vol. 4818, pp. 326–333. Springer, Heidelberg (2008). https://doi.org/10.1007/978-3-540-78827-0_36

21. Filippova, T.F., Matviychuk, O.G.: Reachable sets of impulsive control system with cone constraint on the control and their estimates. In: Lirkov, I., Margenov, S., Wasniewski, J. (eds.) Large-Scale Scientific Computing. LSSC 2011. Lecture Notes in Computer Science, vol. 7116, pp. 123–130. Springer, Berlin, Heidelberg (2012). https://doi.org/10.1007/978-3-642-29843-1_13

22. Filippova, T.F., Matviychuk, O.G.: Estimates of reachable sets of control systems with bilinear-quadratic nonlinearities. Ural Math. J. 1(1), 45–54 (2015)

23. Filippova, T.F.: Estimation of star-shaped reachable sets of nonlinear control systems. In: Lirkov, I., Margenov, S. (eds.) Large-Scale Scientific Computing. LSSC 2017. LNCC, vol. 10665, pp. 210–218. Springer, Cham (2018). https://doi.org/10.1007/978-3-319-73441-5_22

24. Chernousko, F.L., Rokityanskii, D.Y.A.: Ellipsoidal bounds on reachable sets of dynamical systems with matrices subjected to uncertain perturbations. J. Optim. Theory Appl. 104(1), 1–19 (2016)

25. Gusev, M.I.: Application of penalty function method to computation of reachable sets for control systems with state constraints. AIP Conf. Proc. 1773(050003), 1–8 (2016). https://doi.org/10.1063/1.4964973

26. Filippova, T.F.: The HJB approach and state estimation for control systems with uncertainty. IFAC-PapersOnLine. 51(13), 7–12 (2018)

27. Filippova, T.F.: Differential equations for ellipsoidal estimates of reachable sets for a class of control systems with nonlinearity and uncertainty. IFAC-PapersOnLine. 51(32), 770–775 (2018)

28. Kurzhanski, A.B.: Comparison principle for equations of the Hamilton-Jacobi Type. Proc. Steklov Inst. Math. Suppl. 1, 185–195 (2006)

29. Kurzhanski, A.B.: Hamiltonian techniques for the problem of set-membership state estimation. Int. J. Adapt. Control Sig. Process. 25(3), 249–263 (2010)

Structural Instability of Gold and Bimetallic Nanowires Using Monte Carlo Simulation

Vladimir Myasnichenko, Nickolay Sdobnyakov, Leoneed Kirilov, Rossen Mikhov and Stefka Fidanova

Abstract In this paper, we present a method for optimizing of metal nanostructures. The core of the method is a lattice Monte Carlo method with different lattices combined with an approach from molecular dynamics. Interaction between atoms is calculated using multi-body tight-binding model. The method allows solving of problems with periodic boundary conditions. It can be used for modeling of one-dimensional and two-dimensional atomic structures. If periodic boundary conditions are not given, we assume finite dimensions of the model lattice. In addition, automatic relaxation of the crystal lattice can be performed in order to minimize further the potential energy of the system. A computer implementation of the method is developed. It uses the commonly accepted XYZ format for describing atomic structures and passing input parameters. We perform two series of simulations to study the size, composition and temperature dependent surface segregation behaviors and structural atomic instability of Au–Ag nanowires. We found that the most stable mixing configuration of bimetallic nanowires has Ag-rich surface and Au-rich subsurface.

1 Introduction

Metal, including gold, nanowires (filamentary nanocrystals, nanofibers) is a rapidly expanding field of research. Gold nanowires can be used in transparent electrodes for flexible displays [19, 29]. A particularly important point related to electrodes is the stability of nanowires at a thermal load. The minimization of surface energy caused by thermally activated diffusion leads to the rupture of nanowires. This was observed for copper [17], for silver [16], for gold [14], and also for platinum [28].

V. Myasnichenko · N. Sdobnyakov
Tver State University, Tver, Russia
e-mail: viplabs@yandex.ru

L. Kirilov (✉) · R. Mikhov · S. Fidanova
Institute of Information and Communication Technologies, Bulgarian Academy of Sciences,
Sofia, Bulgaria
e-mail: l_kirilov_8@abv.bg

© Springer Nature Switzerland AG 2020
S. Fidanova (ed.), *Recent Advances in Computational Optimization*,
Studies in Computational Intelligence 838,
https://doi.org/10.1007/978-3-030-22723-4_9

The behavior of nanostructures at elevated temperatures can be very different from the macroscopic material. It is well known that small nanoparticles/nanowires will melt at a much lower temperature, which depends on their size [11].

A combination of simulation tools for thermodynamic properties and stability of nanosystems is proposed in [3]; mainly parallel Monte Carlo algorithms for icosahedral, multilayer Pd–Pt clusters. The model is on a 3D cubic lattice. In [5], the chemical ordering in "magic-number" Pd–Ir nanoalloys is studied. The density functional theory is compared with the results of the free energy concentration expansion method. In [8], the problem for stable structures of alloy nanoparticles is investigated. A two-step search strategy is proposed. The first strategy is based on extensive global optimization search and is combined with an empirical potential with density-functional local relaxation. The structure and thermodynamics of Cu–Ni nanoalloys is studied in [25]. An atomic model is described in the framework of a potential based on the second-moment approximation of the tight binding potential. In [26], a novel structure for free Co–Pt nanoalloys is developed. Three computational methodologies have been combined. The energetic stability of the novel structure has been checked.

The structure and energetics of Pd–Pt nanoalloys are studied in [27]. The model is based on the second-moment approximation to tight binding theory. To solve the problem the authors apply a genetic algorithm. In [30] the problem of structure predicting of multi-component systems is studied. The system is represented as a generalized graph. A Kernighan and Lin heuristic procedure is applied to find locally optimal partitions of an arbitrary graph. The same problem is studied in [31]. A local optimization technique combined with multiple local-neighborhood search is applied. In [33] is proposed a parallel modification of the Birmingham cluster genetic algorithm for global optimization of nanoalloy clusters using a pool strategy. The method is illustrated for global optimization of the $Au_{10}Pd_{10}$ cluster using the Gupta potential. The structure of different AuCu clusters is studied in [34] by means of a Parallel Excitable Walkers algorithm and molecular dynamics.

Computer simulation by the molecular dynamics method has been widely used to study the structural defects and the melting temperature of nanowires and gold nanostructures [2, 18] as well as their elasticity and plasticity [7, 38]. The same problems have been studied by means of molecular statics method [24]. In these studies, the significant role of surface tension was determined. The kinetic Monte Carlo was used in [13] for modeling structural transitions and atomic diffusion in gold nanoparticles. In [1] is presented a reliable way to construct a rigid lattice barrier parameterization of face-centered and body-centered cubic metal lattices for the Kinetic Monte Carlo model. Three different barrier sets for Cu and one for Fe are produced that can be used for Kinetic Monte Carlo simulations.

An overview of recent research on bimetallic nanocrystals is presented in [9]. The authors discuss the structural characteristics of the nanocrystals and different approaches for their synthesis with controlled properties. Different applications in industry are also listed. In [21], a liquid-like behavior of gold nanowire bridges is explored. The authors study mechanical and material properties of nanowires for application to nanotechnology. In [23] is demonstrated the transformation of Ag thin films into nanoparticles after single-pulse laser-induced dewetting. The only factor for mean particle size is the initial film thickness. The Rayleigh instability of ultrathin gold nanowires is studied in [37]. The authors use in situ transmission electron microscopy to observe their behavior under real service conditions. In [36] the Rayleigh instability of ultrathin gold nanowires is also studied (diameter < 10 nm). The authors propose simple mechanically assisted self-healing process for their restoring after damaging during practical services.

In [10] the problem for shape instability of nanowires is studied. They use a Kinetic Monte Carlo approach to model the process. A combination of Monte Carlo approach and molecular dynamics is applied for simulating beam induced dynamics of atoms in metallic nanoclusters in [15]. The Kinetic Monte Carlo approach on a coarse-grained lattice is applied in [12] for modeling surface diffusion processes of Ni, Pd and Au nanostructures. It is shown that such a computational approach is able to predict the Rayleigh decay phenomena observed for metallic nanowires with a diameter of several nanometers. The final form after the heating process is reproduced and the calculation scheme correctly predicts the chronological order and location of the nanowire break for any given shape, which can be also checked by TEM-visualization of the heated samples.

One of the important practical applications of the results using Monte Carlo simulations can be seen in the work of [35], where the authors performed a series of experiments on the heating of gold nanowires. It was found that nanowires are especially susceptible to fragmentation around joints and intersection points even at relatively low temperatures. Using the parameterization for Kinetic Monte Carlo, they found that the destruction can be fully explained by the processes of diffusion of atoms, and the destruction of the nanowires will always begin at the junction. The contact point of the nanowire acts as the most preferable place for the diffusion of atoms due to the large number of neighboring atoms present near the intersections of the surface. Thus, the accumulation of atoms leads to the formation of a cluster, which is cut off from the nanowire. Heat treatment allows controlling the speed of the process, which makes it possible to technologically create ordered nanodots that have important applications in electrochemical sensors. Unlike previous work, [22] uses silver nanowires as an example and compares the structure evolution under the influence of nanosecond pulses of laser irradiation and heat treatment in air atmosphere. In both processes, nanowires are initially fragmented into shorter nanorods before being transformed into nanoparticles.

In our opinion, the computer simulation of bimetallic nanosized systems is more difficult in terms of interpreting the results. For example, [32] presents Transmission Electron Microscopy (TEM) studies of nanoscale Ni−Au core−shell particles on heatable TEM grids. But the authors noted that a more representative estimate for the optimal reaction for describing the diffusion of a single Ni atom in an unsupported icosahedral gold cluster ($NiAu_{54}$) would have to be based on statistical approaches such as Monte Carlo sampling.

2   The Monte Carlo Approach

The searching for stable configurations of metal and alloy nanostructures is not a trivial problem from computational point of view. It is equivalent to minimization of the Potential Energy Surface. This function has an extremely large number of local minima. Thus the question for developing efficient and effective methods arises.

The proposed method has four distinctive features. First, a lattice Monte Carlo method with different lattices is used. Second, automatic stretching/compressing of the lattice is done in order to find better optimal solution. Third, Periodic Boundary Conditions (PBCs) are implemented so that the method can be used for modeling of one-dimensional (nanowire, tube) and two-dimensional (nano-film) structures. If periodic boundary conditions are not given, we assume finite dimensions of the model lattice. Fourth, the resulting nanoparticle structures are relaxed at low temperature within molecular dynamics, choosing one of them as an approximation of the global minimum.

Interaction between atoms is calculated using the multi-particle tight-binding potential of Gupta—Cleri and Rosato [4]. The total potential energy of the system is defined as follows:

$$E = \sum_i \left(\sum_{j \neq i} E_{ij}(a, b) - \sqrt{\sum_{j \neq i} B_{ij}(a, b)} \right) \tag{1}$$

$$E_{ij}(a, b) = A_{ab} \exp\left(-p_{ab}\left(\frac{r_{ij}}{r_{0,ab}} - 1 \right) \right) \tag{2}$$

$$B_{ij}(a, b) = \xi_{ab}^2 \exp\left(-2q_{ab}\left(\frac{r_{ij}}{r_{0,ab}} - 1 \right) \right), \tag{3}$$

where i ranges over all atoms; j ranges over all atoms other than i but within distance R_{cut} from i; a and b represent the species of the atoms i and j; $E_{ij}(a, b)$ is the repulsive component of the potential due to the atoms i and j; $B_{ij}(a, b)$ is the binding component of the potential due to the atoms i and j; r_{ij} is the distance between the atoms; $r_{0,ab}$, A_{ab}, p_{ab}, ξ_{ab}, q_{ab} are parameters that depend only on the species of the atoms. R_{cut} is the maximum distance beyond which the interaction is assumed to be zero.

There is no temperature effect increasing the potential energy because comparison of the resulting energy occurs after cooling by the Molecular Dynamics method down to 0.01 K.

According to our method, Periodic Boundary Conditions are defined as follows:

$$r_{ij} = \sqrt{\left|\Delta x_{ij}\right|^2 + |\Delta y_{ij}|^2 + |\Delta z_{ij}|^2}, \, |\Delta x_{ij}| = \min\{|x_i - x_j|, |L_x - |x_i - x_j||\}, \tag{4}$$

where x_i, x_j are the x-coordinates of the atoms i and j inside the periodic cell, and L_x is the size of the periodic cell along the x axis. If the y or z axis is also periodic, then $|\Delta y_{ij}|, |\Delta z_{ij}|$ are also computed in a similar way. The non-periodic case corresponds to $L_x = \infty$.

The complete algorithm consists of the following steps:

Step 1. Read the input data: initial positions of the atoms, the dimensions of the window for Periodic Boundary Conditions, other control parameters, etc. Atoms that do not have their initial positions given are placed at random.

Step 2. For all nodes, pre-compute the lists of neighbors, vicinities and other information that is known ahead of time.

Step 3. Check the exit criteria. The cycle stops when either the requested number of iterations is exceeded, or the system has reached equilibrium. If Yes, go to Step 11.

Step 4. Adjust the temperature according to the following formula:

$$T = \max\{1, T_0 + s\Delta T\}, \tag{5}$$

where T_0 and ΔT are constants, and s is the iteration number. This check is performed once every several thousand iterations.

Step 5. Choose an atom at random.

Step 6. Choose a neighboring empty node at random. If there are no empty neighbors, return to Step 3.

Step 7. Calculate the potential energy difference for the atom moving into the selected empty node, taking into account Periodic Boundary Conditions.

Step 8. If the energy would not increase, perform the jump and return to Step 3.

Step 9. Otherwise, calculate the jump probability $P = \exp(-\Delta E/kT)$ and generate a random number p $(0 \leq p < 1)$.

Step 10. If the number is smaller than the probability, perform the jump, otherwise do nothing. Either way, return to Step 3.

Step 11. Perform iteratively stretching and compressing of the lattice along each axis with step 0.01 (or using another value from an input parameter) to minimize the potential energy further.

3 Computer Implementation

We solve several important issues resulting from the characteristics of the proposed method and from the nature of the problem solved.

We minimize the required computation during the main loop, since it has to be executed for millions of iterations. To this end, we have made extensive use of pre-computation and memoization, inspired by the approach discussed in [20].

A. For each node i, compute the list of nodes j within distance R_{cut} from it. As a sub-list, remember the list of immediate neighbors of i.

B. For each node i, each node j within distance R_{cut} from i, and each combination of a and b, use (2) and (3) to compute the values of $E_{ij}(a, b)$ and $B_{ij}(a, b)$. From this point on, we can forget about Cartesian coordinates and work exclusively with indexes and these pre-computed values using only (1).

C. For each atom i, compute $\sum_{j \neq i (r_{ij} \leq R_{cut})} B_{ij}(a, b)$. This value is remembered and kept updated throughout the algorithm.

These pre-computations allow each step of the main loop to run in constant time. For the energy difference calculation (Step 7), doing it in the naïve way according to (1) would require iterating over not only the R_{cut}-vicinities of the chosen atom and of the chosen empty node, but also over the R_{cut}-vicinities of their R_{cut}-vicinities (since the values under the square roots need to be modified). Pre-computation C allows us to calculate the correct energy by iterating over only the first vicinities.

The data structures are organized as follows. The nodes are kept in two arrays **N** and **A**, where **N** gives the index of a node into **A** and **A** gives the index of a node into **N**. **N** is sorted in input order, while A is sorted to begin with the atoms and end with the empty nodes. This allows us to query information about nodes and atoms, select atoms at random, add, remove and move atoms around, all in constant time.

Input is given as a list of nodes, specified by their Cartesian coordinates. Each node can either be empty or contain an atom. The distance between adjacent nodes in the input data may slightly vary (within 15%). The data format is given in the commonly accepted XYZ format for describing atomic structures.

Other aspects of the problem, such as the number and type of metal atoms, the periodic or non-periodic boundary conditions, the temperature of the system, etc., are specified as additional parameters. These can be given either on the command line, or on the second line ("comment" line) of the input XYZ file.

The output file format is identical to the input file format, with the values for the energy of the system and all other parameters shown on the "comment" line. This allows multiple successive runs of the program on the same data. Output of intermediate results can also be requested.

The program is written in the C programming language and is tested on Windows and Linux platforms. Care has been taken to organize its core functions in a simple way in order to permit compiler optimizations.

Since the proposed method is non-deterministic, several runs of the algorithm are performed on each data set to ensure that the behavior is stable.

4 Numerical Experiments

The method is demonstrated first in the following example: fcc lattice with 6400 nodes of total size 22.6 × 2.1 × 2.3 nm. The X-axis (the longest one, corresponding to the direction of [110]) has periodic boundary conditions. We model 1D structures of 1100, 3000 and 5000 atoms. Two variants of chemical composition are presented:

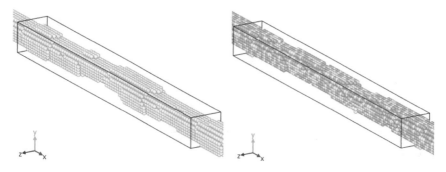

Fig. 1 Au_{3000} (left) and $Au_{1500}Ag_{1500}$ (right) at T = 1 K

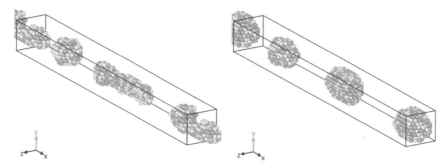

Fig. 2 $Au_{550}Au_{550}$ at T = 1980 K (left) and at T = 1 K (right)

pure gold and gold-silver in a ratio of 1:1. Figure 1 illustrates the configuration for 3000 atoms.

For comparison, 8 different Monte Carlo simulation modes were used: 1 million, 2 million, 4 million, 8 million, 16 million, 32 million, 64 million and 128 million iterations. Simulated annealing was performed cooling down from 2200 to 1 K.

For size 1100 (the thinnest nanowire), a separation into individual nanoparticles of round shape was observed. This is a consequence of the Rayleigh–Plateau instability. Minimizing surface energy (Fig. 2) leads to the breakage of nanowires of both compositions into 3–4 fragments. As will be shown below, for other sizes, the layers/nodes along the Y and Z axes are not enough to obtain this effect.

Figure 3, 4 show the behavior of the algorithm on this data. Figure 3 illustrates the different simulation modes used. Figure 4 gives details about the temperature dependency and jump count for one of the trials as a representative.

A second series of experiments was performed in order to study structure formation in more detail. A bigger fcc lattice with 28,830 nodes was used, increasing in width: 11.9 × 6.1 × 6.1 nm. The X-axis has periodic boundary conditions. Compositions were the same: pure gold and Au–Ag bimetal in a ratio of 1:1. Each was tested with 6 variants for the number of atoms: 1000, 2000, 4000, 6000, 8000 and 16,000. Monte Carlo simulation with 360 million iterations was used, with cooling

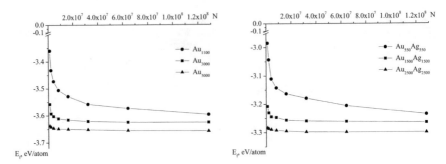

Fig. 3 Dependency of the final potential energy (per atom) on the number of Monte Carlo iterations in 1D gold nanowires (left) and in gold-silver bimetal nanowires (right)

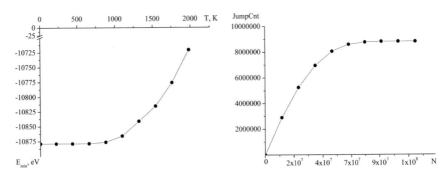

Fig. 4 Au_{3000} cluster cooling in 128 million iterations: dependency of the minimum energy on the temperature (left) and dependency of the number of atom jumps on the iteration number (right)

down from 2200 to 1 K. For each variant, experiments were carried out in series of 30 launches under the same conditions. From each series, the one launch that resulted in the lowest potential energy was selected for structure analysis of its final configuration.

Figure 5 shows the number of single zero-dimensional fragments (clusters) that were observed, separately for gold and bimetal systems. It can be seen that in the case of mono-composition, more atomic clusters are formed, and the average size of each cluster is less than in the case of bimetallic composition. When the occupancy (filling) degree η_{oc} of the nodes reaches 28% in both cases ($N = 8000$), only periodic nanowires are formed.

To get more insight into the structures, it is necessary to analyze the local structural information and atom distribution in the bimetallic nanowires. In [6], concentration oscillations near the surface have been observed layer by layer in $AuAg_3$ and $AuAg$ nanoparticles, where the outmost layer is Ag-rich but the subsurface layer is Au-rich. In our experiments, we observed similar structural characteristics. Although Ag atoms segregate to the surface, the subsurface is mainly Au atoms (60–80%) and the inner layer remains alloy structure.

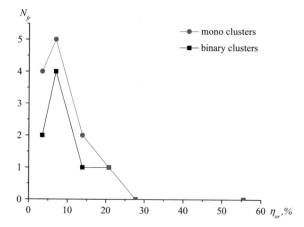

Fig. 5 Dependency of the number of cluster fragments N_{fr} on lattice occupancy rate η_{oc} for mono (Au) and binary (Au–Ag) compositions

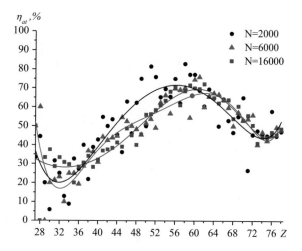

Fig. 6 Comparative dependencies of the fraction/concentration of atoms η_{at} on coordination number Z for gold atoms with different total amount of atoms. Solid curves are polynomial approximations for the corresponding data

The details are shown in Fig. 6. The coordination numbers were calculated for five spheres (instead of the usual one) using a cutoff radius of ~6.8. The surface was considered to consist of the atoms with coordination numbers $Z \leq 50$ and the subsurface—with $51 \leq Z \leq 72$ (see Fig. 7). As can be observed, the segregation effect does not depend on the size (and shape) of the final bimetallic nanostructure.

Cross-section slice snapshots of some of the bimetallic nanostructures with different sizes are shown in Figs. 8 and 9.

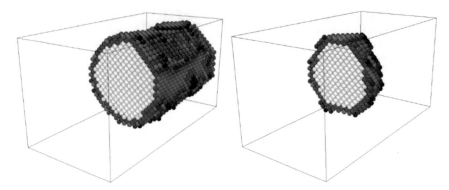

Fig. 7 Cross-section view of the $Au_{8000}Ag_{8000}$ ($N = 16,000$) nanowire (left) and the $Au_{3000}Ag_{3000}$ ($N = 6000$) cluster (right) with color coding: white—core atoms, red—subsurface atoms, gray—surface atoms

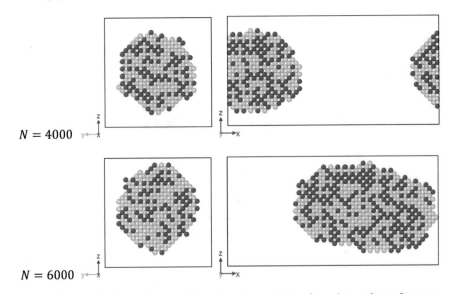

$N = 4000$

$N = 6000$

Fig. 8 Location of gold (yellow) and silver (gray) atoms in the formed nanocluster fragments. Cross-section slices of the 1D lattice by the (100) surface (left) and by the (010) surface (right)

5 Conclusion

A new method for modeling one-dimensional bimetallic nanostructures is proposed. It has four distinctive features: a lattice Monte Carlo method with different lattices is applied; periodic boundary conditions are implemented so that the method can be used for modeling of 1D nanostructures; automatic stretching/compressing of the lattice is done in order to find better optimal solution; the resulting nanowires are relaxed at low temperature within molecular dynamics.

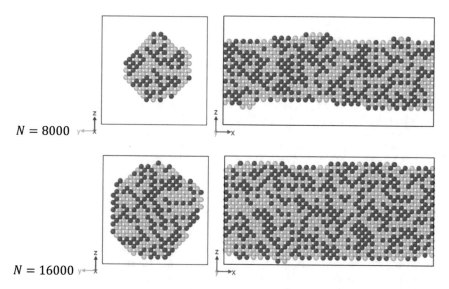

$N = 8000$

$N = 16000$

Fig. 9 Location of gold (yellow) and silver (gray) atoms in the nanowires without fragmentation. Cross-section slices by the (100) surface (left) and by the (010) surface (right)

In this work, we have used this method to study the segregation behavior and atomic-scale structural features of gold and Au–Ag nanowires. Surface segregation of Ag was found in all considered Au–Ag nanowires. The most thermodynamically favorable morphologies are the bimetallic structures with Ag-rich surface, Au-rich subsurface and alloyed core. Separation into individual nanoparticles as a consequence of the Rayleigh–Plateau instability was observed when the lattice filling degree is less than 28%.

Acknowledgements This research is supported by the Russian Foundation for Basic Research project No. 18-38-00571 mol_a and National Scientific Program "Information and Communication Technologies for a Single Digital Market in Science, Education and Security (ICTinSES)", Ministry of Education and Science—Bulgaria and the Bulgarian NSF under the grant DFNI-DN 12/5.

References

1. Baibuz, E., Vigonski, S., Lahtinena, J., Zhao, J., Jansson, V., Zadin, V., Djurabekova, F.: Migration barriers for surface diffusion on a rigid lattice: challenges and solutions. Comput. Mater. Sci. **146**, 287–302 (2018)
2. Bilalbegović, G.: Structures and melting in infinite gold nanowires. Solid State Commun. **115**, 73–76 (2000)
3. Calvo, F.: Solid-solution precursor to melting in onion-ring Pd-Pt nanoclusters: A case of second-order-like phase change? Faraday Discuss. **138**, 75–88 (2008)
4. Cleri, F., Rosato, V.: Tight-binding potentials for transition metals and alloys. Phys. Rev. B **48**, 22–33 (1993)

5. Davis, J., Johnston, R., Rubinovich, L., Polak, M.: Comparative modelling of chemical ordering in palladium-iridium nanoalloys. J. Chem. Phys. **141**, 224307 (2014)
6. Deng, L., Hu, W., Deng, H., Xiao, S., Tang, J.: Au-Ag bimetallic nanoparticles: surface segregation and atomic-scale structure. J. Phys. Chem. C **115**(23), 11355–11363 (2011)
7. Diao, J., Gall, K., Dunn, M.L., Zimmerman, J.A.: Atomistic simulations of the yielding of gold nanowires. Acta Mater. **54**(3), 643–653 (2006)
8. Ferrando, R., Fortunelli, A., Johnston, R.: Searching for the optimum structures of alloy nanoclusters. Phys. Chem. Chem. Phys. **10**, 640–649 (2008)
9. Gilroy, K.D., Ruditskiy, A., Peng, H-Ch., Qin, D., Xia, Y.: Bimetallic nanocrystals: syntheses, properties, and applications. Chem. Rev. **116**(18), 10414–10472 (2016)
10. Gorshkov, V., Privman, V.: Kinetic Monte Carlo model of breakup of nanowires into chains of nanoparticles. J. Appl. Phys. **122**(20), 204301 (2017)
11. Granberg, F., Parviainen, S., Djurabekova, F., Nordlund, K.: Investigation of the thermal stability of Cu nanowires using atomistic simulations. J. Appl. Phys. **115**(21), 213518 (2014)
12. Hausera, A.W., Schnedlitz, M., Ernst, W.E.: A coarse-grained Monte Carlo approach to diffusion processes in metallic nanoparticles. Eur. Phys. J. D **71**, 150 (2017)
13. He, X., Cheng, F., Chen, Z.-X.: The lattice kinetic Monte Carlo simulation of atomic diffusion and structural transformation for gold. Sci. Rep. **6**(1), 33128 (2016)
14. Karim, S., Toimil-Molares, M.E., Balogh, A.G., et al.: Morphological evolution of Au nanowires controlled by Rayleigh instability. Nanotechnology **17**(24), 5954–5959 (2006)
15. Knez, D., Schnedlitz, M., Lasserus, M., Schiffmann, A., Ernst, W.E., Hofer, F.: Modelling electron beam induced dynamics in metallic nanoclusters. Ultramicroscopy **192**, 69–79 (2018)
16. Langley, D.P., Lagrange, M., Giusti, G., Jiménez, C., et al.: Metallic nanowire networks: effects of thermal annealing on electrical resistance. Nanoscale **6**(22), 13535–13543 (2014)
17. Li, H., Biser, J.M., Perkins, J.T., Dutta, S., et al.: Thermal stability of Cu nanowires on a sapphire substrate. J. Appl. Phys. **103**(2), 024315 (2008)
18. Liu, W., Chen, P., Qiu, R., Khan, M., et al.: A molecular dynamics simulation study of irradiation induced defects in gold nanowire. Nucl. Instrum. Methods Phys. Res. Sect. B Beam Interact. with Mater. Atoms. **405**, 22–30 (2017)
19. Luo, M., Liu, Y., Huang, W., Qiao, W.: Towards flexible transparent electrodes based on carbon and metallic materials. Micromachines **8**(1), 12 (2017)
20. Myshlavtsev, A.V., Stishenko, P.V.: Modification of the Metropolis algorithm for modeling metallic nanoparticles. Omsk Sci. Newsp. 1(107), 21–25 (2012). (in Russian)
21. Naik, J., Cheneler, D., Bowen, J., Prewett, P.D.: Liquid-like behaviour of gold nanowire bridges. Appl. Phys. Lett. **111**, 073104 (2017)
22. Oh, H., Lee, J., Lee, M.: Transformation of silver nanowires into nanoparticles by Rayleigh instability: Comparison between laser irradiation and heat treatment. Appl. Surf. Sci. **427**, 65–73 (2018)
23. Oh, Y., Lee, M.: Single-pulse transformation of Ag thin film into nanoparticles via laser-induced dewetting. Appl. Surf. Sci. **399**(31), 555–564 (2017)
24. Olsson, P.A.T., Park, H.S.: Atomistic study of the buckling of gold nanowires. Acta Mater. **59**(10), 3883–3894 (2011)
25. Panizon, E., Olmos-Asar, J., Peressi, M., Ferrando, R.: The study of the structure and thermodynamics of CuNi nanoalloys using a new DFT-fitted atomistic potential. Phys. Chem. Chem. Phys. **17**, 28068–28075 (2015)
26. Parsina, I., DiPaola, C., Baletto, F.: A novel structural motif for free CoPt nanoalloys. Nanoscale **4**, 1160–1166 (2012)
27. Paz-Borbon, L., Mortimer-Jones, T., Johnston, R., Posada-Amarillas, A., et al.: Structures and energetics of 98 atom Pd–Pt nanoalloys: potential stability of the Leary tetrahedron for bimetallic nanoparticles. Phys. Chem. Chem. Phys. **9**, 5202–5208 (2007)
28. Rauber, M., Muench, F., Toimil-Molares, M.E., Ensinger, W.: Thermal stability of electrodeposited platinum nanowires and morphological transformations at elevated temperatures. Nanotechnology **23**(47), 475710 (2012)

29. Sannicolo, T., Lagrange, M., Cabos, A., et al.: Metallic nanowire-based transparent electrodes for next generation flexible devices: a review. Small **12**(44), 6052–6075 (2016)
30. Schebarchov, D., Wales, D.: A new paradigm for structure prediction in multicomponent systems. J Chem Phys. **139**(22), 221101 (2013)
31. Schebarchov, D., Wales, D.: Quasi-combinatorial energy landscapes for nanoalloy structure optimization. Phys. Chem. Chem. Phys. **17**, 28331–28338 (2015)
32. Schnedlitz, M., Lasserus, M., Meyer, R., Knez, D., et al.: Stability of core−shell nanoparticles for catalysis at elevated temperatures: structural inversion in the Ni−Au system observed at atomic resolution. Chem. Mater. **30**, 1113–1120 (2018)
33. Shayeghi, A., Götz, D., Davis, J.B.A., Schäfer, R., Johnston, R.L.: Pool-BCGA: a parallelised generation-free genetic algorithm for the ab initio global optimisation of nanoalloy clusters. Phys. Chem. Chem. Phys. **17**, 2104 (2015)
34. Toai, T.J., Rossi, G., Ferrando, R.: Global optimisation and growth simulation of AuCu clusters. Faraday Discuss. **138**, 49–58 (2008)
35. Vigonski, S., Jansson, V., Vlassov, S., Polyakov, B., et al.: Au nanowire junction breakup through surface atom diffusion. Nanotechnology **29**, 015704 (2018)
36. Wang, B., Han, Y., Xu, Sh, Qiu, L., Ding, F., Lou, J., Lu, Y.: Mechanically assisted self-healing of ultrathin gold nanowires. Small **14**(20), 1704085 (2018)
37. Xu, Sh, Li, P., Lu, Y.: In situ atomic-scale analysis of Rayleigh instability in ultrathin gold nanowires. Nano Res. **11**(2), 625–632 (2018)
38. Zepeda-Ruiz, L.A., Sadigh, B., Biener, J., Hodge, A.M., et al.: Mechanical response of free-standing Au nanopillars under compression. Appl. Phys. Lett. **91**(10), 101907 (2007)

Manipulating Two-Dimensional Animations by Dynamical Distance Geometry

Antonio Mucherino

Abstract The dynamical Distance Geometry Problem (dynDGP) was recently introduced to tackle the problem of manipulating existing animations by modifying and/or adding ad-hoc distance constraints in a distance-based representation of the motion. Although the general problem is NP-hard, satisfactory results have been obtained for the dynDGP by employing local optimization methods, where the original animations, the ones to be manipulated, are given as starting points. New animations are presented in this short paper and, differently from previous publications where only artificial instances were considered, one new animation is extracted from a video clip, depicting animated geometrical objects, that was previously used in a psychological study. The manipulation by distance constraints of such an animation allows to modify the perception of the "actions" performed by the objects of the initial animation.

1 Introduction

Given a positive integer $K > 0$ and a graph $G = (V \times T, E, d)$, the dynamical Distance Geometry Problem (dynDGP) [7] consists in finding a realization

$$x : (v, t) \in V \times T \longrightarrow x_v^t = x(v, t) \in \mathbb{R}^K$$

of G in the Euclidean space $\mathbb{R}^K$ such that the following objective is minimized:

$$\sigma(x) = \sum_{\substack{u,v \in V \\ t,q \in T}} \frac{|\,||x_u^q - x_v^t|| - d(u_q, v_t)\,|}{d(u_q, v_t)}, \tag{1}$$

A. Mucherino (✉)
IRISA, University of Rennes 1, Rennes, France
e-mail: antonio.mucherino@irisa.fr

© Springer Nature Switzerland AG 2020
S. Fidanova (ed.), *Recent Advances in Computational Optimization*,
Studies in Computational Intelligence 838,
https://doi.org/10.1007/978-3-030-22723-4_10

where $|\cdot|$ indicates the absolute value of a real number, and $||\cdot||$ indicates the Euclidean norm.

There are two main differences between the dynDGP and the general DGP [3]. Firstly, while the generic DGP is a decision problem, the dynDGP is generally proposed as an optimization problem, because it is unlikely that all distance constraints related to the general term of $\sigma(x)$ are satisfied in real-life applications; the realizations x where such constraints can be satisfied *as much as possible* are rather searched. Secondly, in the dynDGP, the vertex set is the set product between two sets, a set V of *objects*, and a set T representing the time as a sequence of discrete steps. As a consequence, the vertex (v, t) of the graph is an ordered pair that represents a given object v at a certain instant t, and a realization x corresponds to an animation of the objects in V over the time steps in T. As in the equations above, the compact notations v_t and $\{u_q, v_t\}$ are employed in this paper for indicating a vertex and an edge of G, respectively. The position $x(v, t)$ of the object v at the time t is indicated with x_v^t.

In the dynDGP, a second real value can be associated to every edge of the graph: the priority $\pi(u_q, v_t)$ on every available edge weight $d(u_q, v_t)$. This additional value can be exploited by solution methods for assigning a different importance to the distance constraints to be satisfied. In Eq. (1), in order to give a different importance to different distances accordingly with the value $\pi(u_q, v_t)$, this value needs to be multiplied by the corresponding term in the general sum in $\sigma(x)$. For more details, the reader is referred to [4].

The class of dynDGP instances that are considered in this short paper are constructed from known initial animations, given by the trajectories x_v^t of the objects $v \in V$ for all times $t \in T$. In order to create dynDGP instances, such initial animations are represented by the relative distances between a subset of vertices pairs, and they are then manipulated by modifying or including new distance constraints. In this way, new animations, having some particular desired properties, can be obtained. The set of obtained distances defines a new instance of the dynDGP, which is represented by a graph G where priorities can be associated to every distance. Since newly added constraints are in general a minority wrt the ones deduced from the original animation, they generally need to have a larger priority π. Although the general DGP is an NP-hard problem [3], satisfactory results have been found for the dynDGP with local optimization methods. When the new introduced constraints are not "too much" in conflict with the initial animation, these initial animations are in fact good starting points for local optimization methods.

The attention in this paper is given to animations of independent objects, i.e. the focus is on animations where the objects induce, for no times t of T, any subgraph with skeletal structure (the reader interested in animations with skeletal structures is referred to [6]). Animations of independent objects can be found, for example, in the context of crowd simulations, and aircraft and multi-robot systems (see [7] and citations therein).

Since the initial animations can serve as a good starting point for dynDGP solutions, a Spectral Projected Gradient (SPG) algorithm, coupled with a non-monotone line-search, was recently exploited for this kind of problems [5]. Gradient descent methods are typically used for local optimization, where the direction given by the vector opposite to the objective gradient is explored for minimizing the function values [1].

In this short paper, the manipulation of two kinds of animations, in dimension $K = 2$, is presented. In Sect. 2, an artificial animation of two moving objects, which intersect at certain times t, is modified so that collisions are avoided while keeping as much as possible the original trajectories of the two objects. In Sect. 3, an animation, created for a psychology study published in 1944 by Heider and Simmel, is extracted from the original video clip and manipulated in order to emphasize the perception of the object's apparent behavior. The dynDGP solver is written in C, while the program for extracting an animation from a given video clip is written in Java (in the experiments presented in this paper, both programs were run under Linux). Finally, Sect. 4 will conclude the paper with some final remarks.

2 A Sinusoidal Animation

This animation is artificially constructed in the 2-dimensional Euclidean space by generating the trajectories of two objects as follows:

- the x coordinates change constantly with the time t (both objects);
- the y coordinates are computed as $10 \cos(x)$ (object 1) or $10 \sin(x)$ (object 2).

The animation takes place in the real box region $[0, 2\pi] \times [-10, 10]$ and is composed by 100 frames. From the initial trajectories, it is possible to obtain a representation that is solely based on inter-object distances, by considering either distances between the two objects at the same time, or distances between two positions of the same object at two different times. However, in order to define the graph G (the dynDGP instance), it is necessary to identify the distances that better describe the objects' motion: as discussed in [7], such distances are the inter-frame distances between the objects at different times t. In the presented experiments, the distances between every object and its own position k frames earlier is considered, with k ranging from 1 to 3. Moreover, in order to avoid collisions occurring in the initial animation, it is imposed that all distances between the objects at the same time t are greater than a predefined positive real threshold Δ. It is important to point out that the animations are not treated frame by frame by the optimization method, but rather as a whole.

Figure 1 shows the solutions (the animations) obtained with the non-monotone SPG [5], for different values of the threshold Δ. Different degrees of blue (from dark blue to light blue) are used for showing the temporal component of the animation (see gray scale in the print version). In the original animation in Fig. 1a, the degree of blue of the two objects is the same at the point where the two curves intersect (see the bottom side of the curve): the two objects collide in fact in the original animation. Figure 1b shows the solution found by solving the dynDGP instance where $\Delta = 0.1$.

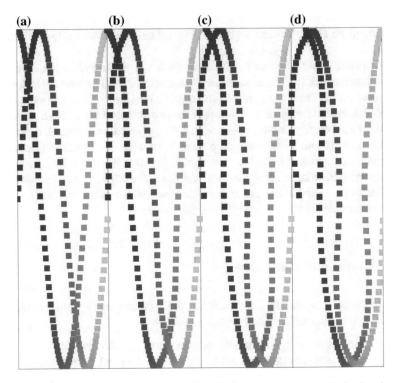

Fig. 1 The trajectories of two objects with a sinusoidal trajectory: **a** the original animation (see degrees of blue for temporal component); **b** dynDGP solution with $\Delta = 0.1$; **c** dynDGP solution with $\Delta = 0.2$; **d** dynDGP solution with $\Delta = 0.3$

The two trajectories are slightly modified, so that the collisions can be avoided (the two objects cannot be closer than 0.1 in the overall animation), while preserving as much as possible the original animation. This modification is more and more evident in the solutions obtained with $\Delta = 0.2$ (see Fig. 1c) and $\Delta = 0.3$ (see Fig. 1d).

3   Heider and Simmel Animation

In 1944, Heider and Simmel published an original psychological study about perception, where panelists were asked to express the behavior they perceived while viewing a video clip depicting a set of animated objects. Such objects had, for the first time in this study, no human qualities, but they were rather represented with geometrical figures such as triangles (one larger and another smaller) and one circle, together with third geometrical figure representing the "house". The house was the only inanimated object, unless it was "pushed" by the other objects. The performed experiments showed that most people interpreted the movements of the geometrical objects as actions of animated beings (in most cases persons). Some of panelists were even able to "see" short stories out from the object's animation [2].

The trajectories of the moving objects were extracted from part of the original video clip[1] (only the frames where the two triangles and the circle are all present were considered, that is, the first 100 frames of the video clip were not taken into consideration). Differently from the experiments in the previous section, the objects' shapes, together with the local orientation of such shapes (in particular for the two triangles) are extracted from the original animation, but left unchanged during the distance-based manipulations. As in the experiments in Sect. 2, dynDGP instances are created from these original trajectories, while including some ad-hoc distance constraints in order to add some new visual effects in obtained solutions. In total, the video clip is composed by 1800 frames; the "house" was not considered in our experiments.

As in the experiments presented in the previous section, the first dynDGP instance related to Heider and Simmel's video clip is generated with the idea of avoiding contacts among the objects "taking part" to the scene. In the particular context of the present video, this reflects the fact that the objects should not come too close to each other. Starting from frame 250, the viewer sees in the original animation the two triangular objects attacking each other by suddenly approaching to one another. In the dynDGP solution, instead, this approaching behavior is not pronounced, because of the imposed distance constraints (with $\Delta = 0.2$, where the environment is the real box $[0, 1] \times [0, 1]$). However, the attacking effect is a way amplified with this modification. In fact, in the obtained dynDGP solution, it is not necessary for one triangle to approach "too much" the other to make it step back; a much lighter approach is actually sufficient to have the stepping back effect over the other. In Fig. 2, this new obtained animation is referred to as `solution 1`.

A second experiment was performed on this video clip where, instead of avoiding collisions, a close approach of the objects was encouraged. Given a threshold Δ (fixed to 0.2 as in the previous experiment), the dynDGP instance is now created in a way that, every time the distance between two objects is larger than Δ in the original animation, the interval $[\varepsilon, \Delta]$ is imposed for the distance, where ε is a small real number (set to 0.005 in this experiment). The animation obtained as a solution of the generated dynDGP instance gives a much stronger feeling of stress and attack in comparison with the original animation created by Heider and Simmel, even at the very initial frames (in the original animation, the "attacks" start around frame 250). This new obtained animation is referred to as `solution 2`.

The video clips generated from the found solutions are available online,[2] while Fig. 2 compares some key frames of the original animation with the two found dynDGP solutions.

---

[1] https://www.youtube.com/watch?v=sx7lBzHH7c8.

[2] https://www.antoniomucherino.it/en/animations.php.

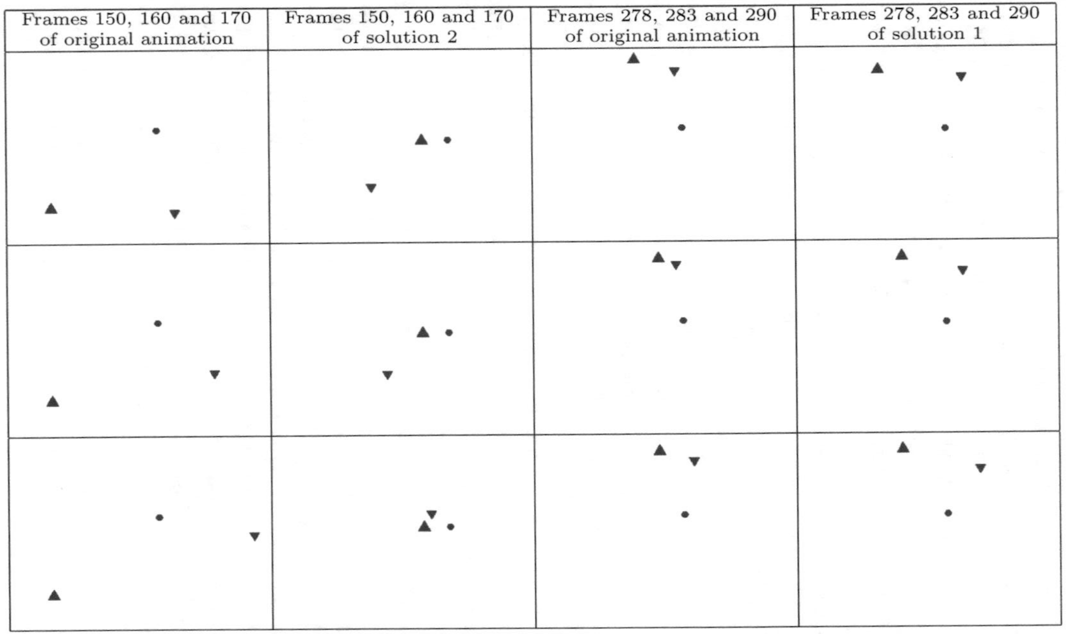

Fig. 2 Comparisons between some key frames in the original animation and the obtained dynDGP solutions. In the first two columns, solution 2 is compared with the original animation at frames 150, 160 and 170: the original open configuration does not exist anymore and this gives a sense of agitation to the overall animation. In the last two columns, solution 1 is compared with the original animation at frames 278 (the two triangles stare at each other), 283 (the big triangle attacks) and 290 (the small triangle makes a step back): the same behavior is visible in the new obtained animation but without a close approach of the two objects

4 Conclusions

The dynDGP is a new research direction that is attracting more and more attention in recent times [5, 7, 8]. In this short paper, an original experiment was proposed, where an animation was extracted from a video clip initially used in a psychological study, and manipulated so that the perception of the actions by the viewer are emphasized. Although the presented results are already interesting, research is necessary in a near future to validate the overall approach, and to make it more general and robust.

References

1. Glunt, W., Hayden, T.L., Raydan, M.: Molecular conformations from distance matrices. J. Comput. Chem. **14**(1), 114–120 (1993)
2. Heider, F., Simmel, M.: An experimental study of apparent behavior. Am. J. Psychol. **57**(2), 243–259 (1944)
3. Liberti, L., Lavor, C., Maculan, N., Mucherino, A.: Euclidean distance geometry and applications. SIAM Rev. **56**(1), 3–69 (2014)
4. Mucherino, A.: On the Discretization of Distance Geometry: Theory, Algorithms and Applications, HDR Monograph, University of Rennes 1. INRIA Hal archive id: tel-01846262, 17 July 2018
5. Mucherino, A., Gonçalves, D.S.: An Approach to Dynamical Distance Geometry. Lecture Notes in Computer Science, vol. 10589; Nielsen, F., Barbaresco, F. (eds.): Proceedings of Geometric Science of Information (GSI17), Paris, France, pp. 821–829 (2017)
6. Mucherino, A., Gonçalves, D.S., Bernardin, A., Hoyet, L., Multon, F.: A distance-based approach for human posture simulations. In: IEEE Conference Proceedings, Federated Conference on Computer Science and Information Systems (FedCSIS17), Workshop on Computational Optimization (WCO17), Prague, Czech Republic, pp. 441–444 (2017)
7. Mucherino, A., Omer, J., Hoyet, L., Robuffo Giordano, P., Multon, F.: An application-based characterization of dynamical distance geometry problems. Optim. Lett. (2019) (Springer) (to appear)
8. Tabaghi, P., Dokmanić, I., Vetterli, M.: Kinetic Euclidean Distance Matrices, 13 pp. (2018). arXiv preprint arXiv:1811.03193

Index Matrices as a Cost Optimization Tool of Resource Provisioning in Uncertain Cloud Computing Environment

Velichka Traneva, Stoyan Tranev and Vassia Atanassova

Abstract The main objective in cloud computing environment is efficient allocation of its resources in order to get optimal overall cost (time) for doing tasks. In today's market environment, the time or costs for doing tasks are not exact numbers, often provided data about their values may be incomplete. The intuitionistic fuzzy sets are developed to deal with these weaknesses of conventional crisp and fuzzy sets. The problem for allocating resources in uncerain cloud environment efficiently in order to get optimal cost is an intuitionistic fuzzy assignment problem. In the paper a new type of three-dimensional assignment problem (3D-IFAP) is formulated, in which the costs of assigning tasks to resources are intuitionistic fuzzy pairs and the priority of tasks is known in advance. We propose intuitionistic fuzzy Hungarian algorithm, based on the concepts of index matrices and intuitionistic fuzzy logic, to optimally solve this problem.

1 Introduction

Cloud computing is "a model for enabling convenient, on-demand network access to a shared pool configurable computing resources that can be rapidly provisioned and released with minimal management effort or service provider interaction" [40]. The

This work was supported by the Bulgarian Ministry of Education and Science under the National Research Programme "Young scientists and postdoctoral students", approved by DCM # 577/ 17.08.2018.

V. Traneva (✉) · S. Tranev
"Prof. Asen Zlatarov" University,"Prof. Yakimov" Blvd, 8000 Bourgas, Bulgaria
e-mail: veleka13@gmail.com

S. Tranev
e-mail: tranev@abv.bg

V. Atanassova
Institute of Biophysics and Biomedical Engineering, Bulgarian Academy of Sciences, Sofia, Bulgaria
e-mail: vassia.atanassova@gmail.com

main objective of cloud environment is optimal assignment activities to resources with minimal cost (time for completion or for response).

Various methods for optimizing cloud environments have been proposed. A credit based algorithm for solving task scheduling problem in cloud computing environment in order to obtain optimal time is described in [50]. Some algorithms for placing virtual machines efficiently in cloud environment in order to minimize cost for accessing these machines have been proposed in [1, 23, 31, 60]. In [48] the use of particle swarm optimization algorithm has been introduced for scheduling jobs over a set of resources and has been compared with the best resource selection algorithm with the objective to minimize the computational cost. In [19] cat swarm optimization was applied for minimizing computational cost in a distributed cloud environment, by providing an effective mapping scheme for workflow scheduling. The paper [59] presents an application of the Bat Algorithm for scheduling a workflow application. In [20] the autors have modified the bee colony optimization algorithm to make it suitable for task scheduling on cloud systems. In [68] the firefly algorithm has been implemented for workflow scheduling.

Some autors have proposed the problem for the cost optimization of resource provisioning in cloud computing to be modelled as an assignment problem (AP). The AP was first introduced by Stone in [67] and is known to be NP-hard [21]. The basic polynomial time algorithm is the Hungarian method, which was developed by Kuhn in 1955 [32] and solves the problem for $O(n^4)$ time, where n is the number of sourses or tasks that need to be assigned. The method is subsequently reviewed by Munkres [45], who offers the reduction of the worst-case computational complexity from $O(n^4)$ to $O(n^3)$. Bertsekas [18] proposed an efficient implementation of the shortest path algorithm and showed that the algorithm has a theoretical complexity of $O(n^3)$. Balas [16] represented the parallel implementation of the shortest path algorithm for assignment problems. A faster version of the assignment problem is also proposed in Wright [73], which implements the minor modification in cost matrix to reduce the execution time.

The three-dimensional assignment problem (3D-AP) is an extension of the two-dimensional AP. Algorithms for the 3D-AP have been developed. An algorithm to solve 3D-AP, which indirectly evaluates all feasible solutions, was given by Pierskalla [51]. Olivo [47] suggested an algorithm that mixes the Hungarian method [33, 34] for solving AP and the branch and bound technique given in [39] to solve the travelling salesman problem. In [27] was shown that a certain 3D-AP is NP-complete. Hung and Lim [29] proposed a local search genetic algorithm-based method to solve the 3D-AP, but they could not guarantee that an optimal solution could be obtained. 3D-AP model was presented in [38] and two branch and bound based algorithms were proposed to solve the model.

The modern dynamic environment predetermines the frequent changes in parameters in the market economy [15]. The data about assignment parameters of the cost optimization problem in cloud computing environment may be incomplete, leading to vagueness and lack of precision. The intuitionistic fuzzy logic proposed by Atanassov [2] as an extension of Zadeh's fuzzy logic [74] provides us with tools for modeling in the uncertain environment. Many researchers have studied AP in a fuzzy environment. In [35, 46] a fuzzy assignment problem (FAP) is defined, which is first converted into classical problem by Robust ranking method and then Hungarian method is used to solve it. The row penalty method for solving fuzzy assignment problem was discussed in [55]. A parallel moving method is used for solving FAP in [49]. The optimal solution of FAP is obtained using fuzzy Hungarian algorithm in [52]. Fuzzy algorithms of APs are defined in [36, 37, 44, 63], a quadratic FAP is researched in [42], multi-criteria FAPs are given in [17, 28], two-objective FAPs are defined in [24, 28], and multi-job fuzzy APs in [41].

Intuitionistic fuzzy assignment problem (IFAP) with the trapezoidal intuitionistic fuzzy numbers is solved in [53]. In [54] was proposed an algorithm finding an optimal solution of IFAP with triangular intuitionistic fuzzy costs (special case of intuitionistic fuzzy numbers). Intuitionistic fuzzy Hungarian method, based on the intuitionistic fuzzy sets and index matrices concepts, is proposed in [72] for finding an optimal solution of IFAP. In [22] the autors identified a set of relevant decision criteria and their subcriteria needed in the evaluation of the cloud computing technology provider selection problem in uncertain environment.

In the paper, which is an extension of [72], is modelled by the concepts of intuitionistic fuzzy logic and index matrices the resource assignment process in uncertain cloud computing environment and propose three-dimensional intuitinistic fuzzy Hungarian algorithm to find the optimal distribution of tasks to resources over time (by location or other). The costs/times for doing tasks by virtual machines are intuitionistic fuzzy pairs (IFPs) and also are known in advance as the priority of tasks.

The concept of IMs was introduced in 1984 in [2, 3]. IMs appear as an auxiliary tool for describing different mathematical objects, but they are mainly used to describe the transitions in generalized nets [2, 4, 7, 56, 62, 64, 65, 75], intuitionistic fuzzy relations and graphs with finite sets of vertices [5, 6, 8–10] as well as some algorithms for decision making [8, 14, 25, 26, 43, 57, 58, 66].

The rest of this paper is structured as follows: Sect. 2 describes the related concepts of the index matrices (IMs) and IFPs. In Sect. 3, we propose an algorithm for three-dimensional assignment problem (3D-IFAP) with intuitionistic fuzzy costs of doing of tasks, based on the Hungarian algorithm, in uncertain cloud environment by using the concepts of index matrices (IMs) and intuitionistic fuzzy sets (IFSs). The proposed example in Sect. 4 demonstrates the correctness and efficiency of the algorithm. Section 5 offers the conclusion and outlines some aspects for future research.

2 Basic Definitions

This section provides some remarks on IFPs (see [8, 13]) and on index matrices (see [11, 69, 71]).

2.1 Short Notes on Intuitionistic Fuzzy Pairs

The **IFP** is an object of the form $\langle a, b \rangle = \langle \mu(p), \nu(p) \rangle$, where $a, b \in [0, 1]$ and $a + b \leq 1$, that is used as an evaluation of a proposition p [12, 13]. $\mu(p)$ and $\nu(p)$ respectively determine the "truth degree" and "falsity degree".

Let us have two IFPs $x = \langle a, b \rangle$ and $y = \langle c, d \rangle$. In [13] are defined the following operations

$$\neg x = \langle b, a \rangle; \qquad\qquad x \wedge_1 y = \langle \min(a, c), \max(b, d) \rangle;$$
$$x \vee_1 y = \langle \max(a, c)), \min(b, d) \rangle; \qquad x \wedge_2 y = x + y = \langle a + c - a.c, b.d \rangle;$$
$$x \vee_2 y = x.y = \langle a.c, b + d - b.d \rangle; \qquad \alpha.x = \langle 1 - (1 - a)^\alpha, b^\alpha \rangle (\alpha \in R)$$
$$x - y \quad = \langle \max(0, a - c), \min(1, b + d, 1 - a + c) \rangle$$

and relations

$$x \geq y \text{ iff } a \geq c \text{ and } b \leq d; \ x \leq y \text{ iff } a \leq c \text{ and } b \geq d; \tag{1}$$
$$x = y \text{ iff } a = c \text{ and } b = d.$$

The IFP x is an **"intuitionistic fuzzy false pair"** (IFFP) [72] if and only if $a \leq b$, while x is a **"false pair"** (FP) iff $a = 0, b = 1$.

Let a set E be fixed. An **"intuitionistic fuzzy set"** (IFS) A in E is an object of the following form (see [8]):

$$A = \{\langle x, \mu_A(x), \nu_A(x) \rangle | x \in E\},$$

where functions $\mu_A : E \to [0, 1]$ and $\nu_A : E \to [0, 1]$ define the degrees of membership and non-membership of x, respectively, and for every $x \in E$:

$$0 \leq \mu_A(x) + \nu_A(x) \leq 1.$$

2.2 Definition of 3D-IFIM, Operations and Relations

Let $\mathscr{I}$ be a fixed set. By three-dimensional intuitionistic fuzzy index matrix (3D-IFIM) with index sets K, L and H $(K, L, H \subset \mathscr{I})$, we denote the object:

$$[K, L, H, \{\langle \mu_{k_i, l_j, h_g}, \nu_{k_i, l_j, h_g} \rangle\}]$$

$$\equiv \begin{array}{c|cccccc} h_g \in H & l_1 & \cdots & l_j & \cdots & l_n \\ \hline k_1 & \langle \mu_{k_1, l_1, h_g}, \nu_{k_1, l_1, h_g} \rangle & \cdots & \langle \mu_{k_1, l_j, h_g}, \nu_{k_1, l_j, h_g} \rangle & \cdots & \langle \mu_{k_1, l_n, h_g}, \nu_{k_1, l_n, h_g} \rangle \\ \vdots & \vdots & \ddots & \vdots & \ddots & \vdots \\ k_m & \langle \mu_{k_m, l_1, h_g}, \nu_{k_m, l_1, h_g} \rangle & \cdots & \langle \mu_{k_m, l_j, h_g}, \nu_{k_m, l_j, h_g} \rangle & \cdots & \langle \mu_{k_m, l_n, h_g}, \nu_{k_m, l_n, h_g} \rangle \end{array},$$

where for every $1 \leq i \leq m, 1 \leq j \leq n, 1 \leq g \leq f$:

$$0 \leq \mu_{k_i,l_j,h_g}, \nu_{k_i,l_j,h_g}, \mu_{k_i,l_j,h_g} + \nu_{k_i,l_j,h_g} \leq 1.$$

Following [11, 69], we recall some operations over two IMs
$A = [K, L, H, \{\langle \mu_{k_i,l_j,h_g}, \nu_{k_i,l_j,h_g} \rangle\}]$ and $B = [P, Q, E, \{\langle \rho_{p_r,q_s,e_d}, \sigma_{p_r,q_s,e_d} \rangle\}]$:

Negation

$$\neg A = [K, L, H, \{\langle \nu_{k_i,l_j,h_g}, \mu_{k_i,l_j,h_g} \rangle\}].$$

Addition-(max,min)

$$A \oplus_{(\max,\min)} B = [K \cup P, L \cup Q, H \cup E, \{\langle \phi_{t_u,v_w,x_y}, \psi_{t_u,v_w,x_y} \rangle\}],$$

where

$$\langle \phi_{t_u,v_w,x_y}, \psi_{t_u,v_w,x_y} \rangle$$

$$= \begin{cases} \langle \mu_{k_i,l_j,h_g}, \nu_{k_i,l_j,h_g} \rangle, & \text{if } t_u = k_i \in K, v_w = l_j, x_y = h_g \in H - E \\ & \text{or } t_u = k_i \in K, v_w = l_j \in L - Q, x_y = h_g \in H \\ & \text{or } t_u = k_i \in K - P, v_w = l_j \in L, x_y = h_g \in H \\[2mm] \langle \rho_{p_r,q_s,e_d}, \sigma_{p_r,q_s,e_d} \rangle, & \text{if } t_u = p_r \in P, v_w = q_s \in Q, x_y = e_d \in E - H \\ & \text{or } t_u = p_r \in P, v_w = q_s \in Q - L, x_y = e_d \in E \\ & \text{or } t_u = p_r \in P - K, v_w = q_s \in Q, x_y = e_d \in E \\[2mm] \langle \max(\mu_{k_i,l_j,h_g}, \rho_{p_r,q_s,e_d}), & \text{if } t_u = k_i = p_r \in K \cap P, v_w = l_j = q_s \in L \cap Q \\ \quad \min(\nu_{k_i,l_j,h_g}, \sigma_{p_r,q_s,e_d}) \rangle, & \text{and } x_y = h_g = e_d \in H \cap E \\ \langle 0, 1 \rangle, & \text{otherwise} \end{cases}$$

The definition of the operations $A \oplus_{(\min,\min)} B$ and $A \oplus_{(\min,\max)} B$ are similar.

Termwise multiplication- (min, max)

$$A \otimes_{(\min,\max)} B = [K \cap P, L \cap Q, H \cap R, \{\langle \phi_{t_u,v_w,x_y}, \psi_{t_u,v_w,x_y} \rangle\}],$$

where $\langle \phi_{t_u,v_w,x_y}, \psi_{t_u,v_w,x_y} \rangle = \langle \min(\mu_{k_i,l_j,h_g}, \rho_{p_r,q_s,e_d}), \max(\nu_{k_i,l_j,h_g}, \sigma_{p_r,q_s,e_d}) \rangle$.

The definition of the operations $A \otimes_{(\min,\min)} B$ and $A \otimes_{(\min,\max)} B$ are similar.

Reduction: We use symbol "$\perp$" for lack of some component in the separate definitions. In some cases, it is suitable to change this symbol with "0". The operation $(k, \perp, \perp)$-reduction of a given IM A is defined by:

$$A_{(k,\perp,\perp)} = [K - \{k\}, L, H, \langle \phi_{t_u,v_w,x_y}, \psi_{t_u,v_w,x_y} \rangle],$$

where

$$\langle \phi_{t_u,v_w,x_y}, \psi_{t_u,v_w,x_y} \rangle = \langle \mu_{k_i,l_j,h_g}, \nu_{k_i,l_j,h_g} \rangle \text{ for } t_u = k_i \in K - \{k\}, v_w = l_j \in L \text{ and}$$

$$x_y = h_g \in H.$$

The definitions of the $(\perp, l, \perp)$- and $(\perp, \perp, h)$-reductions are analogous.

Projection: Let $M \subseteq K$, $N \subseteq L$ and $U \subseteq H$. Then,

$$pr_{M,N,U}A = [M, N, U, \{b_{k_i,l_j,h_g}\}],$$

where for each $k_i \in M$, $l_j \in N$ and $h_g \in U$, $b_{k_i,l_j,h_g} = a_{k_i,l_j,h_g}$.

Substitution: Let the IM $A = [K, L, H, \{a_{k,l,h}\}]$ be given. Local substitution over A is defined for the pair of indices (p, k) by

$$\left[\frac{p}{k}; \perp; \perp\right]A = \left[(K - \{k\}) \cup \{p\}, L, H, \{a_{k,l,h}\}\right].$$

The definitions of operations $\left[\perp; \frac{q}{l}; \perp\right]A$ and $\left[\perp; \perp; \frac{e}{h}\right]A$, where (q, l) and (e, h) are pairs of indices, are analogous.

Index type operations:

$$AGIndex_{(\min,\max),(\not{\perp})}(A) = \langle k_i, l_j, h_g \rangle,$$

which finds the index of the minimum element of A that has no empty value.

If $AGIndex_{(\min,\max),(\not{\perp})}(A)$ can not be determined, then

$$AGIndex_{(\min,\max),(\not{\perp})}(A) = \perp.$$

$$Index_{(\not{\perp})/(1)}(A) = \{\langle k_{i_1}, l_{j_1}, h_{g_1} \rangle, \ldots, \langle k_{i_x}, l_{j_x}, h_{g_x} \rangle, \ldots, \langle k_{i_u}, l_{j_u}, h_{g_u} \rangle\},$$

where $\langle k_{i_x}, l_{j_x}, h_{g_x} \rangle$ (for $1 \le x \le u$) is the index of the element of A, whose cell is full (or the value 1).

Let us define the operations in a similar way to that in [72]

$$Index_{(\max/\min \mu/v), k_i, h_g}(A) = \{\langle k_i, l_{v_1}, h_g \rangle, \ldots, \langle k_i, l_{v_x}, h_g \rangle, \ldots, \langle k_i, l_{v_V}, h_g \rangle\},$$

where $\langle k_i, l_{v_x}, h_g \rangle$ (for $1 \le i \le m$, $1 \le v \le n$, $1 \le x \le V$, $1 \le g \le f$) is the indices of the IFFP of $k_i \in K$ of A for a fixed $h_g \in H$, for which v_{k_i,l_{v_x},h_g} or v_{k_i,l_{v_x},h_g} is maximum/minimum.

$$Index_{(\max/\min \mu/v), l_j, h_g}(A) = \{\langle k_{w_1}, l_j, h_g \rangle, \ldots, \langle k_{w_y}, l_j, h_g \rangle, \ldots, \langle k_{w_W}, l_j, h_g \rangle\},$$

where $\langle k_{w_y}, l_j, h_g \rangle$ (for $1 \le j \le n$, $1 \le g \le f$, $1 \le y \le W$, $1 \le w \le m$) are the indices of the IFFP of $l_j \in L$ for a fixed $h_g \in H$ of A, for which $\mu_{k_{w_y},l_j,h_g}$ or $v_{k_{w_y},l_j,h_g}$ is maximum/minimum.

$$Index_{(\max/\min \mu/v), h_g, k_i}(A) = \{\langle k_i, l_{v_1}, h_g \rangle, \ldots, \langle k_i, l_{v_x}, h_g \rangle, \ldots, \langle k_i, l_{v_V}, h_g \rangle\},$$

where $\langle k_i, l_{v_x}, h_g \rangle$ (for $1 \leq i \leq m, 1 \leq g \leq f, 1 \leq v \leq n, 1 \leq x \leq V$) are the indices of the IFFP of $h_g \in H$ for a fixed $k_i \in K$ of A, for which μ_{k_i,l_{v_x},h_g} or ν_{k_i,l_{v_x},h_g} is maximum/minimum.

$$Index_{(\max / \min \mu / \nu), h_g, l_j}(A) = \{\langle k_{w_1}, l_j, h_g \rangle, \ldots, \langle k_{w_y}, l_j, h_g \rangle, \ldots, \langle k_{w_W}, l_j, h_g \rangle\},$$

where $\langle k_{w_y}, l_j, h_g \rangle$ (for $1 \leq i \leq m, 1 \leq g \leq f, 1 \leq w \leq m, 1 \leq y \leq W$) are the indices of the IFFP of $h_g \in H$ for fixed $l_j \in L$ of A, for which $\mu_{k_{w_y},l_j,h_g}$ or $\nu_{k_{w_y},l_j,h_g}$ is maximum/minimum.

Aggregate global internal operation:

$$AGIO_{\oplus_{(\max,\min)}}(A)$$

This operation finds the "$\oplus_{(\max,\min)}$"-operation of all the matrix elements.

Internal subtraction of IMs' components [70, 71]:

$$IO_{-(\min,\max)}(\langle k_i, l_j, h_g, A \rangle, \langle p_r, q_s, e_d, B \rangle) = [K, L, H, \{\langle \gamma_{t_u,v_w,z_y}, \delta_{t_u,v_w,z_y} \rangle\}],$$

where $k_i \in K$, $l_j \in L$, $h_g \in H$; $p_r \in P$, $q_s \in Q, e_d \in E$ and

$$\langle \gamma_{t_u,v_w,z_y}, \delta_{t_u,v_w,z_y} \rangle$$

$$
= \begin{cases}
\langle \mu_{t_u,v_w,z_y}, \nu_{t_u,v_w,z_y} \rangle, & \text{if } t_u \neq k_i \in K, v_w \neq l_j \in L, \\
& z_y \neq h_g \in H \\
\langle \max(0, \mu_{k_i,l_j,h_g} - \rho_{p_r,q_s,e_d}), & \text{if } t_u = k_i \in K, \\
\min(1, \nu_{k_i,l_j,h_g} + \sigma_{p_r,q_s,e_d}, 1 - \mu_{k_i,l_j,h_g} + \rho_{p_r,q_s,e_d}) \rangle, & v_w = l_j \in L, z_y = h_g \in H.
\end{cases}
$$

The non-strict relation "inclusion about value" is defined by:

$$A \subseteq_v B \text{ iff } (K = P)\&(L = Q)\&(H = R)\&(\forall k \in K)(\forall l \in L)(\forall h \in H)$$

$$(a_{k,l,h} \leq b_{k,l,h}).$$

3 Intuitionistic Fuzzy Hungarian Method for Optimization of Resource Provisioning Cost in Cloud Computing

In this section, we solve the three-dimensional form of an intuitionistic fuzzy assignment problem (3D-IFAP) to allocate cloudlets to virtual machines by the Hungarian algorithm, based on IMs concept. Its formulation is:

m virtual machines (VMs) $\{k_1, \ldots, k_i, \ldots, k_m\}$ need to be assigned for n cloudlets $\{l_1, \ldots, l_j, \ldots, l_n\}$ in a time-moment h_g ($g = 1, \ldots, f$). H is a fixed time scale and h_g is its element. The interpretation of H may also be different such as location or other. The cost of assigning c_{k_i,l_j,h_g} of the l_jth task to the k_ith machine at time (or place) h_g is an intuitionistic fuzzy cost. Task priority is defined in advance.

The objective of the 3D-IFAP is to minimize the total cost of completing all jobs of the virtual machines over time so that each task can be done by only one machine at a time.

The mathematical model of the problem is as follows:

$$\text{minimize} \sum_{i=1}^{m} \sum_{j=1}^{n} \sum_{g=1}^{f} c_{k_i,l_j,h_g} x_{k_i,l_j,h_g}$$

$$\text{Subject to: } \sum_{j=1}^{n} \sum_{g=1}^{f} x_{k_i,l_j,h_g} = 1, \qquad i = 1, 2, \ldots, m;$$

$$\sum_{i=1}^{m} \sum_{g=1}^{f} x_{k_i,l_j,h_g} = \langle 1, 0 \rangle, \qquad j = 1, 2, \ldots, n; \tag{2}$$

$$\sum_{i=1}^{m} \sum_{j=1}^{n} x_{k_i,l_j,h_g} = \langle 1, 0 \rangle, \qquad g = 1, 2, \ldots, f;$$

$$x_{k_i,l_j,h_g} \in \{\langle 0, 1 \rangle \text{ or } \langle 1, 0 \rangle\}, \quad \text{for} \quad 1 \le i \le m, \quad 1 \le j \le n, \quad 1 \le g \le f$$

The additional constraint on task priorities is given by a 3D-IFIM "*PREF*".

Note: The operations "addition" and "multiplication", used in the problem (2) are those for IFPs, defined in Sect. 2.1.

Let us construct the following 3D-IFIM, in accordance with the problem (2):

$$C[K, L, H]$$

$$= \quad \begin{array}{c|cccc} h_g \in H & l_1 & \cdots & l_n & R \\ \hline k_1 & \langle \mu_{k_1,l_1,h_g}, \nu_{k_1,l_1,h_g} \rangle & \cdots & \langle \mu_{k_1,l_n,h_g}, \nu_{k_1,l_n,h_g} \rangle & \langle \mu_{k_1,R,h_g}, \nu_{k_1,R,h_g} \rangle \\ \vdots & \vdots & \ddots & \vdots & \vdots \\ k_m & \langle \mu_{k_m,l_1,h_g}, \nu_{k_m,l_1,h_g} \rangle & \cdots & \langle \mu_{k_m,l_n,h_g}, \nu_{k_m,l_n,h_g} \rangle & \langle \mu_{k_m,R,h_g}, \nu_{k_m,R,h_g} \rangle \\ Q & \langle \mu_{Q,l_1,h_g}, \nu_{Q,l_1,h_g} \rangle & \cdots & \langle \mu_{Q,l_n,h_g}, \nu_{Q,l_n,h_g} \rangle & \langle \mu_{Q,R,h_g}, \nu_{Q,R,h_g} \rangle \end{array}, \tag{3}$$

where $K = \{k_1, k_2, \ldots, k_m, Q\}$, $L = \{l_1, l_2, \ldots, l_n, R\}$, $H = \{h_1, h_2, \ldots, h_f, T\}$ and for $1 \le i \le m, 1 \le j \le n, 1 \le g \le f$: $\{c_{k_i,l_j,h_g}, c_{k_i,R,h_g}, c_{Q,l_j,h_g}, c_{k_i,l_j,T}\}$ are IFPs. Let us denote by $|K| = m + 1$ the number of elements of the set K; then $|L| = n + 1$ and $|H| = f + 1$. We also define

$$X[K*, L*, H*, \{x_{k_i,l_j,h_g}\}] = \quad \begin{array}{c|ccccc} h_g \in H* & l_1 & \cdots & l_j & \cdots & l_n \\ \hline k_1 & x_{k_1,l_1,h_g} & \cdots & x_{k_1,l_j,h_g} & \cdots & x_{k_1,l_n,h_g} \\ \vdots & \vdots & \ddots & \vdots & \ddots & \vdots \\ k_m & x_{k_m,l_1,h_g} & \cdots & x_{k_m,l_j,h_g} & \cdots & x_{k_m,l_n,h_g} \end{array},$$

where $K* = \{k_1, k_2, \ldots, k_m\}$, $L* = \{l_1, l_2, \ldots, l_n\}$, $H* = \{h_1, h_2, \ldots, h_f\}$, and for $1 \leq i \leq m, 1 \leq j \leq n, 1 \leq g \leq f$:

$$x_{k_i, l_j, h_g} = \langle \rho_{k_i, l_j, h_g}, \sigma_{k_i, l_j, h_g} \rangle = \begin{cases} \langle 1, 0 \rangle, & \text{if VM } k_i \text{ is assigned to the task } l_j \\ & \quad \text{in a time-moment } h_g \\ \langle 0, 1 \rangle, & \text{otherwise} \end{cases}$$

and let in the begining of the algorithm x_{k_i, l_j, h_g} is an empty cell for every $k_i \in K*$, $l_j \in L*$, $h_g \in H*$.

Additionally, it is necessary to create the auxiliary IM of task priorities:

$$P[k_1, L*, H*, \{p_{k_1, l_j, h_g}\}],$$

where $L* = \{l_1, l_2, \ldots, l_n\}$, $H* = \{h_1, h_2, \ldots, h_f\}$ and $p_{k_1, l_j, h_g} = \langle \delta_{k_1, l_j, h_g}, \gamma_{k_1, l_j, h_g} \rangle$ is the priority of the l_jth activity in a time-moment h_g.

We compose the IM of task priorities

$$PREF[K*, L*, H*, \{pref_{k_i, l_j, h_g}\}]$$

$$= P \oplus_{(max,min)} \cdots \oplus_{(max,min)} \begin{bmatrix} k_i \\ k_1 \end{bmatrix}; \perp; \perp \end{bmatrix} P \oplus_{(max,min)} \cdots \oplus_{(max,min)} \begin{bmatrix} k_m \\ k_1 \end{bmatrix}; \perp; \perp \end{bmatrix} P$$

where $K* = \{k_1, k_2, \ldots, k_m\}$, $L* = \{l_1, l_2, \ldots, l_n\}$, $H* = \{h_1, h_2, \ldots, h_f\}$ and $k_i \in K*$, $l_j \in L*$ and $h_g \in H*$.

Let us define the following auxiliary matrices:

$$S = [K, L, H, \{s_{k_i, l_j, h_g}\}],$$

such that $S = C$, i.e., $(s_{k_i, l_j, h_g} = c_{k_i, l_j, h_g} \; \forall k_i \in K, \forall l_j \in L, \forall h_g \in H)$;

$$D[K*, L*, H*]$$

where $K* = \{k_1, k_2, \ldots, k_m\}$, $L* = \{l_1, l_2, \ldots, l_n\}$, $H* = \{h_1, h_2, \ldots, h_f\}$, and for $1 \leq i \leq m, 1 \leq j \leq n, 1 \leq g \leq f$: $d_{k_i, l_j, h_g} \in \{1, 2, 3\}$, if the element s_{k_i, l_j, h_g} of S is crossed out with 1, 2 or 3 lines respectively;

$$RC[K*, e_0] = \begin{array}{c|c} & e_0 \\ \hline k_1 & rc_{k_1, e_0} \\ \vdots & \vdots \\ k_m & rc_{k_m, e_0} \end{array},$$

where $K* = \{k_1, k_2, \ldots, k_m\}$ and for $1 \leq i \leq m$: rc_{k_i, e_0} is equal to 0 or 1, depending on whether the k_ith index of $K*$ of the matrix S is crossed out;

$$CC[r_0, L*] = \frac{\begin{array}{cccccc} l_1 & \cdots & l_j & \cdots & l_n \end{array}}{r_0 \begin{array}{cccccc} cc_{r_0,l_1} & \cdots & cc_{r_0,l_j} & \cdots & cc_{r_0,l_n} \end{array}},$$

where $L* = \{l_1, l_2, \ldots, l_n\}$ and for $1 \leq j \leq n$: $cc_{k_i,l_j}\}$ is equal to 0 or 1, depending on whether the l_jth index of $L*$ of the matrix S is crossed out;

$$TC[H*, t_0] = \begin{array}{c|c} & t_0 \\ \hline h_1 & tc_{h_1,t_0} \\ \vdots & \vdots \\ h_g & tc_{h_g,t_0} \\ \vdots & \vdots \\ h_f & tc_{h_f,t_0} \end{array},$$

where $H* = \{h_1, h_2, \ldots, h_f\}$ and for $1 \leq g \leq f$: $tc_{h_g,t_0} = \{0 \text{ or } 1\}$ depending on whether the h_gth index of $H*$ of the matrix S is crossed out.

Let in the begining

$$rc_{k_i,e_0} = cc_{r_0,l_j} = tc_{h_g,t_0} = 0, x_{k_i,l_j,h_g} = \langle 0, 1 \rangle (\forall k_i \in K*, l_j \in L*, h_g \in H*).$$

We will propose a new approach to the Hungarian algorithm [32] for finding the optimal solution of the assignment problem (2) with intuitionistic fuzzy data with the tools of IMs and IFPs. A part of Microsoft Visual Studio.NET 2010 C project's program code [77] is used in the algorithm. The software "IndMatCalc v 0.9" [76] is used to implement the matrix operations.

Step 1. We compare the number of elements of the sets K, L and H in C.

Step 1.1. If the number of elements of K is greater than the number of elements of L, then a dummy index $l_{n+1} \in \mathscr{I}$ is entered in the C, in which all prices c_{k_i,l_{n+1},h_g} $(i = 1, \ldots, m; g = 1, \ldots, f)$ are equal to $\langle 1, 0 \rangle$; otherwise, go to *Step 1.2*. For this purpose, the following operations are executed:

– we define the 3D-IM $C_1[K/\{Q\}, \{l_{n+1}\}, H/\{T\}, \{c_{1_{k_i,l_{n+1},h_g}}\}]$, whose elements are equal to $\langle 1, 0 \rangle$;

– the new cost matrix C is obtained by:
$C := C \oplus_{(\max,\min)} C_1$; $s_{k_i,l_j,h_g} = c_{k_i,l_j,h_g}$, $\forall k_i \in K, \forall l_j \in L, \forall h_g \in H$;
Go to *Step 2*.

Step 1.2. If the number of elements of L is greater than the number of elements of K, then a dummy index $k_{m+1} \in \mathscr{I}$ is entered in the matrix C, in which all prices c_{k_{m+1},l_j,h_g} $(j = 1, \ldots, n; g = 1, \ldots, f)$ are equal to $\langle 1, 0 \rangle$; otherwise, go to *Step 1.3*. Similar operations to those in *Step 1.1.* are executed.

Step 1.3. If the number of elements of K is greater than the number of elements of H, then a dummy index $h_{f+1} \in \mathscr{I}$ is entered in the matrix C, in which all prices $c_{k_i,l_j,h_{f+1}}$ $(i = 1, \ldots, m; j = 1, \ldots, n)$ are equal to $\langle 1, 0 \rangle$; otherwise, go to the *Step 1.4*.

Similar operations to those in *Step 1.1.* by IMs are executed.

Step 1.4.. If the number of elements of H is greater than the number of elements of H, then a dummy index $k_{m+1} \in \mathscr{I}$ is entered in the matrix C, in which all prices c_{k_{m+1},l_j,h_g} $(j = 1, \ldots, n; g = 1, \ldots, f)$ are equal to $\langle 1, 0 \rangle$; otherwise, go to the *Step 2*. The similar operations to those in *Step 1.1.* are executed.

Step 2. Let us create the IM DC, such that

$$DC := C_{(\{Q\},\{R\},\{T\})} \otimes_{(\min,\max)} \neg PREF;$$

$$C := pr_{K,R,H} C \oplus_{(\max,\min)} pr_{Q,L,H} C \oplus_{(\max,\min)} pr_{Q,R,T} C \oplus_{(\max,\min)} DC.$$

Let us create IM $S = [K, L, \{s_{k_i,l_j,h_g}\}]$ such that $S = C$, i.e., $s_{k_i,l_j,h_g} = c_{k_i,l_j,h_g}$ $\forall k_i \in K$, $\forall l_j \in L$ and $\forall h_g \in H$.

Step 3. For each index k_i of K at a fixed h_g $(g = 1, \ldots, f)$ of the matrix S, the smallest element is found among the elements s_{k_i,l_j,h_g} $(j = 1, \ldots, n)$ and it is subtracted from all elements $s_{k_i,l_j,h_g}, j = 1, 2, \ldots, n$. We get the matrix S with reduced prices. Go to *Step 4*.

Step 3.1. For each index k_i of K at a fixed h_g $(g = 1, \ldots, f)$ of the matrix S, the smallest element is found and is recorded as the value of the element s_{k_i,R,h_g}:

for (int $= 0; g < f; g + +$)
for (int $i = 0; i < m; i + +$)
for (int $j = 0; j < n; j + +$)
$\{AGIndex_{(\min,\max),(\perp)} \left(pr_{k_i,L,h_g} S \right) = \langle k_i, l_{v_j}, h_g \rangle;$
If $AGIndex_{(\min,\max),(\perp)} \left(pr_{k_i,L,h_g} S \right) = \perp$ then execute the operation

$$Index_{(\max \mu),k_i,h_g} \left(pr_{k_i,L,h_g} S \right);$$

We create IMs S_1 *and* S_2 :

$$S_1[k_i, l_{v_j}, h_g] = pr_{k_i,l_{v_j},h_g} S;$$

$$S_2 = \left[\perp; \frac{R}{l_{v_j}}; \perp \right] S1;$$

$$S := S \oplus_{(\max,\min)} S_2.\}$$

Step 3.2. For each index k_i of K of the matrix S at a fixed h_g $(g = 1, \ldots, f)$, we subtract the smallest element s_{k_i,l_{v_j},h_g} from the elements s_{k_i,l_j,h_g} $(j = 1, \ldots, n)$.

We create IM $B = pr_{K,R,H} S$.

for (int $g = 0; g < f; g + +$)
for (int $i = 0; i < m; i + +$)
for (int $j = 0; j < m; j + +$)
If $s_{k_i,l_j,h_g} \neq \perp$, then $\{IO_{-(\max,\min)} \left(\langle k_i, l_j, h_g, S \rangle, \langle k_i, R, h_g, B \rangle \right)\}$.

Step 4. For each index l_j of L at a fixed h_g $(g = 1, \ldots, f)$ of the matrix S, the smallest element is found among the elements s_{k_i, l_j, h_g} $(i = 1, \ldots, m)$ and it is subtracted from all elements s_{k_i, l_j, h_g}, for $i = 1, 2, \ldots, m$. We get the matrix S with reduced costs and go to *Step 5*.

Step 4.1. For each index l_j of L at a given h_g $(g = 1, \ldots, f)$ of the matrix S, the smallest element is found and is recorded as the value of the element s_{Q, l_j, h_g}:

for (int $g = 0$; $g < f$; $g + +$)
for (int $j = 0$; $j < n$; $j + +$)

$$\{AGIndex_{(\min, \max),(\perp)} \left(pr_{K, l_j, h_g} S \right) = \langle k_{w_i}, l_j, h_g \rangle;$$

If $AGIndex_{(\min, \max),(\perp)} \left(pr_{K, l_j, h_g} S \right) = \perp$ then execute the operation

$$Index_{(\max \mu), l_j, h_g} \left(pr_{K, l_j, h_g} S \right);$$

Let us create IMs S_3 and S_4 :

$$S_3[k_{w_i}, l_j, h_g] = pr_{k_{w_i}, l_j, h_g} S; \quad S_4 = \left[\frac{Q}{k_{w_i}}; \perp; \perp \right] S3;$$

$$S := S \oplus_{(\max, \min)} S_4.\}$$

Step 4.2. For each index l_j of L of the matrix S at a given $h_g (g = 1, \ldots, f)$, we subtract the smallest element $s_{k_{w_i}, l_j, h_g}$ from all elements s_{k_i, l_j, h_g} $(i = 1, \ldots, m)$.

We create IM $B = pr_{Q, L, h_g} S$.

for (int $j = 0$; $i < n$; $j + +$)
for (int $g = 0$; $g < f$; $g + +$)
for (int $i = 0$; $i < m$; $i + +$)
If $s_{k_i, l_j, h_g} \neq \perp$, then $\{IO_{-(\min, \max)} \left(\langle k_i, l_j, h_g, S \rangle, \langle Q, l_j, h_g, B \rangle \right)\}$.

Step 5. For each index k_i of K at a fixed l_j $(j = 1, \ldots, n)$ of the matrix S, the smallest element is found among the elements s_{k_i, l_j, h_g} $(g = 1, \ldots, f)$ and it is subtracted from all elements s_{k_i, l_j, h_g}, for $g = 1, 2, \ldots, f$ and go to *Step 6*.

Step 5.1. For each index k_i of K at a fixed l_j $(j = 1, \ldots, n)$ of the IM S, the smallest element is found and is recorded as the value of the element $s_{k_i, l_j, T}$:

for (int $i = 0$; $i < m$; $i + +$)
for (int $j = 0$; $j < n$; $j + +$)

$$\{AGIndex_{(\min, \max),(\perp)} \left(pr_{k_i, l_j, H} S \right) = \langle k_i, l_j, h_{z_g} \rangle;$$

If $AGIndex_{(\min, \max),(\perp)} \left(pr_{k_i, l_j, H} S \right) = \perp$, then execute the operation

$$Index_{(\max \mu), k_i, l_j} \left(pr_{k_i, l_j, H} S \right);$$

Let us create IMs S_5 and S_6 :

$$S_5[k_i, l_j, h_{z_g}] = pr_{k_i, l_j, h_{z_g}} S; \ S_6 = \left[\perp; \perp; \frac{T}{h_{z_g}} \right] S_5;$$

$$S := S \oplus_{(+)} S_6\}.$$

Step 5.2. For each index k_i of K of S at a fixed l_j $(j = 1, \ldots, n)$ we subtract the smallest element $s_{k_i, l_j, h_{z_g}}$ from all elements s_{k_i, l_j, h_g} $(j = 1, \ldots, f)$.

We create IM $B = pr_{K, l_j, T} S$.

for (int $i = 0; i < m; i + +$)
for (int $j = 0; j < n; j + +$)
for (int $g = 0; g < f; g + +$)
If $s_{k_i, l_j, h_g} \neq \perp$, then

$$\{IO_{-(\min, \max)} \left(\langle k_i, l_j, h_g, S \rangle, \langle k_i, l_j, T, B \rangle \right)\}.$$

Step 6. Cross out all elements $\langle 0, 1 \rangle$ in the matrix S with the minimum possible number of lines (horizontal, vertical or both). If the number of these lines is $m.f$, go to *Step 8*. If the number of lines is less than $m.f$, go to *Step 7*.

This step introduces IM $D[K*, L*, H*]$, which has the same structure as the X matrix. We use to mark whether an element in S is crossed out with a line to the direction of the dimension K, L or H, or with all three types.

- If $d[k_i, l_j, h_g] = 1$, then the element $s[k_i, l_j, h_g]$ is covered with one line;
- if $d[k_i, l_j, h_g] = 2$, then the element $s[k_i, l_j, h_g]$ is covered with two lines;
- if $d[k_i, l_j, h_g] = 3$, then the element $s[k_i, l_j, h_g]$ is covered with three lines.

We create three IMs $CC[r_0, L*]$, $RC[K*, e_0]$ and $TC[H*, t_0]$, in which is recorded that some index k_i $(i = 1, \ldots, m)$ of K, or l_j $(j = 1, \ldots, n)$ of L, or h_g $(g = 1, \ldots, f)$ of H is covered with a line in the matrix S.

for (int $i = 0; i < m; i + +$)
for (int $j = 0; j < n; j + +$)
for (int $g = 0; g < f; g + +$)

- If $s[k_i, l_j, h_g] = \langle 0, 1 \rangle$ (or $\langle k_i, l_j, h_g \rangle \in Index_{(\max v), k_i, h_g}(S)$) and $d[k_i, l_j, h_g] = 0$, then

$$\{rc[k_i, e_0] = 1;$$

$$\text{for (int } i = 0; i < m; i + +) \ d[k_i, l_j, h_g] = 1;$$

$$S_{(k_i, \perp, \perp)}\}.$$

- If $s[k_i, l_j, h_g] = \langle 0, 1 \rangle$ (or $\langle k_i, l_j \rangle \in Index_{(\max v), k_i, h_g}(S)$) and $d[k_i, l_j, h_g] = 1$, then

$$\{d[k_i, l_j, h_g] = 2; \ cc[r_0, l_j] = 1;$$

$$\text{for (int } j = 0; j < n; j + +) \, d[k_i, l_j, h_g] = 1;$$

$$S_{(\perp, l_j, \perp)}\}.$$

- If $s[k_i, l_j, h_g] = \langle 0, 1 \rangle$ and $d[k_i, l_j, h_g] = 2$, then

$$\{d[k_i, l_j, h_g] = 3; \, tc[h_g, t_0] = 1;$$

$$\text{for (int } g = 0; g < f; g + +) \, d[k_i, l_j, h_g] = 1;$$

$$S_{(\perp, \perp, h_g)}\}.$$

Then we count the number of lines in each of the three dimensions K, L and H, i.e., the count of elements whose value is 1 in the IMs CC, RC and TC.

$$Index_{(1)} (RC) = \{\langle k_{u_1}, e_0 \rangle, \ldots, \langle k_{u_i}, e_0 \rangle, \ldots, \langle k_{u_x}, e_0 \rangle\};$$

$$Index_{(1)} (CC) = \{\langle r_0, l_{v_1} \rangle, \ldots, \langle r_0, l_{v_j} \rangle, \ldots, \langle r_0, l_{v_y} \rangle\};$$

$$Index_{(1)} (TC) = \{\langle h_{w_1}, t_0 \rangle, \ldots, \langle h_{w_g}, t_0 \rangle, \ldots, \langle h_{w_z}, t_0 \rangle\}.$$

If count($Index_{(1)} (RC)$)+count($Index_{(1)} (CC)$)+ count($Index_{(1)} (TC)$) $= m.f = n.f$, then go to *Step 8.*, otherwise to *Step 7*.

Step 7. We find the smallest element in the IM S that it is not crossed by the lines in *Step 6.*, and subtract it from any uncovered element of S, and we add it to each element, which is covered by three lines. We return to *Step 6.* with the resulting reduced price matrix S.

The operation

$$AGIndex_{(\min, \max)} (S) = \langle k_x, l_y, h_z \rangle$$

finds the smallest element index among the elements of the S matrix.

The operation subtract it from any uncovered element of S

$$IO_{-(\max, \min)} \left(\langle S \rangle, \langle k_x, l_y, h_z, S \rangle \right).$$

We add it to each element of S, which is crossed out by three lines:
for (int $g = 0; g < f; g + +$)
for (int $i = 0; i < m; i + +$)
for (int $j = 0; j < n; j + +$)
{if $d[k_i, l_j, h_g] = 3$, then

$$S_1 = pr_{k_x, l_y, h_z} C;$$

$$S_2 = pr_{k_i,l_j,h_g} C \oplus_{(\max,\min)} \left[\frac{k_i}{k_x}; \frac{l_j}{l_y}; \frac{h_g}{h_z} \right] S_1;$$

$$S := S \oplus_{(\max,\min)} S_2;$$

if $d[k_i, l_j, h_g] = 1$ or $d[k_i, l_j, h_g] = 2$ then

$$S := S \oplus_{(\max,\min)} pr_{k_i,l_j,h_g} C\};$$

Go to *Step 6*.

Step 8. An optimal solution has been found, where assignments of the VMs to the tasks in a time-moment are located where the elements $\langle 1, 0 \rangle$ in the X, with a single such element in each dimension.
{for (int $i = 0; i < m; i++$)
for (int $j = 0; j < n; j++$)
for (int $g = 0; g < f; g++$)
if $(\langle k_i, l_j, h_g \rangle \in Index_{(\max v),k_i}(S))$ and $d[k_i, l_j, h_g] \neq 4$), then

$$x[k_i, l_j, h_g] = \langle 1, 0 \rangle;$$

for (int $i = 0; i < m; i++$) $d[k_i, l_j, h_g] = 4;$
for (int $j = 0; j < n; j++$) $d[k_i, l_j, h_g] = 4;$
for (int $g = 0; g < f; g++$) $d[k_i, l_j, h_g] = 4\}.$

The optimal solution is in $X_{opt}[K*, L*, H * \{x_{k_i,l_j,h_g}\}]$ and the optimal assignment cost is:

$$AGIO_{\oplus_{(\max,\min)}} \left(C_{(\{Q\},\{R\},\{T\})} \otimes_{(\min,\max)} X_{opt} \right).$$

4 An Example of the Application of the 3D-Intiutionistic Fuzzy Hungarian Algorithm by IMs for Cost Optimization of Cloud Computing Environment

Let us consider the following problem as an application of the algorithm, presented in Sect. 3:

Three virtual machines (VMs) $\{k_1, k_2, k_3\}$ need to be assigned for four cloudlets $\{l_1, l_2, l_3\}$ in a time-moment h_g ($g = 1, 2, 3$). H is a fixed time scale and h_g is its element. The cost of assigning c_{k_i,l_j,h_g} of l_jth cloudlet to k_ith VMs at a time h_g is an intuitionistic fuzzy cost.

Cloudlet priorities are defined in advance by 3D-IFIM "*PREF*". The cost for asigning c_{k_i,l_j} of the k_ith candidate for the l_jth job is an IFP and is defined as an element of the 2D-IFIM C. The purpose of the problem is to minimize the total cost of assigning cloudlets to VMs.

Solution of the problem: The software "IndMatCalc v 0.9" [76] was used to implement the matrix operations.

Let us be given the following IMs, in accordance with the problem:

$$C[K, L, H] = \left\{ \begin{array}{c|cccc} h_1 & l_1 & l_2 & l_3 & R \\ \hline k_1 & \langle 0.8, 0.1 \rangle & \langle 0.7, 0.1 \rangle & \langle 0.6, 0.4 \rangle & \langle \perp, \perp \rangle \\ k_2 & \langle 0.5, 0.4 \rangle & \langle 0.8, 0.1 \rangle & \langle 0.8, 0.2 \rangle & \langle \perp, \perp \rangle \\ k_3 & \langle 0.7, 0.3 \rangle & \langle 0.7, 0.2 \rangle & \langle 0.5, 0.3 \rangle & \langle \perp, \perp \rangle \\ Q & \langle \perp, \perp \rangle & \langle \perp, \perp \rangle & \langle \perp, \perp \rangle & \langle \perp, \perp \rangle \end{array} \right. ,$$

$$\begin{array}{c|cccc} h_2 & l_1 & l_2 & l_3 & R \\ \hline k_1 & \langle 0.7, 0.2 \rangle & \langle 0.5, 0.2 \rangle & \langle 0.6, 0.1 \rangle & \langle \perp, \perp \rangle \\ k_2 & \langle 0.6, 0.1 \rangle & \langle 0.7, 0.1 \rangle & \langle 0.5, 0.4 \rangle & \langle \perp, \perp \rangle \\ k_3 & \langle 0.7, 0.2 \rangle & \langle 0.5, 0.3 \rangle & \langle 0.6, 0.2 \rangle & \langle \perp, \perp \rangle \\ Q & \langle \perp, \perp \rangle & \langle \perp, \perp \rangle & \langle \perp, \perp \rangle & \langle \perp, \perp \rangle \end{array} ,$$

$$\begin{array}{c|cccc} h_3 & l_1 & l_2 & l_3 & R \\ \hline k_1 & \langle 0.8, 0.1 \rangle & \langle 0.6, 0.2 \rangle & \langle 0.7, 0.1 \rangle & \langle \perp, \perp \rangle \\ k_2 & \langle 0.7, 0.1 \rangle & \langle 0.6, 0.3 \rangle & \langle 0.6, 0.3 \rangle & \langle \perp, \perp \rangle \\ k_3 & \langle 0.6, 0.3 \rangle & \langle 0.5, 0.3 \rangle & \langle 0.7, 0.2 \rangle & \langle \perp, \perp \rangle \\ Q & \langle \perp, \perp \rangle & \langle \perp, \perp \rangle & \langle \perp, \perp \rangle & \langle \perp, \perp \rangle \end{array} ,$$

$$\left. \begin{array}{c|cccc} T & l_1 & l_2 & l_3 & R \\ \hline k_1 & \langle \perp, \perp \rangle & \langle \perp, \perp \rangle & \langle \perp, \perp \rangle & \langle \perp, \perp \rangle \\ k_2 & \langle \perp, \perp \rangle & \langle \perp, \perp \rangle & \langle \perp, \perp \rangle & \langle \perp, \perp \rangle \\ k_3 & \langle \perp, \perp \rangle & \langle \perp, \perp \rangle & \langle \perp, \perp \rangle & \langle \perp, \perp \rangle \\ Q & \langle \perp, \perp \rangle & \langle \perp, \perp \rangle & \langle \perp, \perp \rangle & \langle \perp, \perp \rangle \end{array} \right\} ,$$

where $K = \{k_1, k_2, k_3, Q\}$, $L = \{l_1, l_2, l_3, R\}$, $H = \{h_1, h_2, h_3, T\}$ and for $1 \leq i \leq 3$, $1 \leq j \leq 3, 1 \leq g \leq 3 : \{c_{k_i, l_j, h_g}, c_{k_i, R, h_g}, c_{Q, l_j, h_g}, c_{k_i, l_j, T}\}$ are IFPs.

We can see that $|K| = |L| = |H| = 3$ and the problem is balanced. We also define

$$X[K*, L*, H*] = \left\{ \begin{array}{c|ccc} h_g \in H & l_1 & l_2 & l_3 \\ \hline k_1 & \langle 0, 1 \rangle & \langle 0, 1 \rangle & \langle 0, 1 \rangle \\ k_2 & \langle 0, 1 \rangle & \langle 0, 1 \rangle & \langle 0, 1 \rangle \\ k_3 & \langle 0, 1 \rangle & \langle 0, 1 \rangle & \langle 0, 1 \rangle \end{array} \right\} ,$$

$K* = \{k_1, k_2, k_3\}$, $L* = \{l_1, l_2, l_3\}$ and $H* = \{h_1, h_2, h_3\}$.

The IM of preferences is

$$PREF[K*, L*, H*] = \left\{ \begin{array}{c|ccc} h_1 & l_1 & l_2 & l_3 \\ \hline k_1 & \langle 0.5, 0.4 \rangle & \langle 0.7, 0, 2 \rangle & \langle 0.8, 0, 1 \rangle \\ k_2 & \langle 0.5, 0.4 \rangle & \langle 0.7, 0, 2 \rangle & \langle 0.8, 0, 1 \rangle \\ k_3 & \langle 0.5, 0.4 \rangle & \langle 0.7, 0, 2 \rangle & \langle 0.8, 0, 1 \rangle \end{array} \right. '$$

$$\begin{array}{c|ccc} h_2 & l_1 & l_2 & l_3 \\ \hline k_1 & \langle 0.5, 0.3 \rangle & \langle 0.6, 0, 3 \rangle & \langle 0.7, 0, 2 \rangle \\ k_2 & \langle 0.5, 0.3 \rangle & \langle 0.6, 0, 3 \rangle & \langle 0.7, 0, 2 \rangle \\ k_3 & \langle 0.5, 0.3 \rangle & \langle 0.6, 0, 3 \rangle & \langle 0.7, 0, 2 \rangle \end{array} '\quad \begin{array}{c|ccc} h_3 & l_1 & l_2 & l_3 \\ \hline k_1 & \langle 0.7, 0.3 \rangle & \langle 0.6, 0, 3 \rangle & \langle 0.9, 0, 1 \rangle \\ k_2 & \langle 0.7, 0.3 \rangle & \langle 0.6, 0, 3 \rangle & \langle 0.9, 0, 1 \rangle \\ k_3 & \langle 0.7, 0.3 \rangle & \langle 0.6, 0, 3 \rangle & \langle 0.9, 0, 1 \rangle \end{array} \right\} .$$

Step 1. We compare $|K|$, $|L|$ and $|H|$. The problem is balanced and go to *Step 2*.
Step 2. We create IM DC, such that

$$DC := C_{(\{Q\},\{R\},\{T\})} \otimes_{(\min,\max)} \neg PREF;$$

$$C := pr_{K,R,H} C \oplus_{(\max,\min)} pr_{Q,L,H} C \oplus_{(\max,\min)} pr_{Q,R,T} C \oplus_{(\max,\min)} DC.$$

Let us create the IM $S = [K, L, \{s_{k_i, l_j, h_g}\}]$, such that $S = C$, i.e., $s_{k_i, l_j, h_g} = c_{k_i, l_j, h_g}$
$\forall k_i \in K$, $\forall l_j \in L$ and $\forall h_g \in H$.
So, the matrix $C = S$ acquires the following form:

$$S[K, L, H] = \left\{ \begin{array}{c|cccc} h_1 & l_1 & l_2 & l_3 & R \\ \hline k_1 & \langle 0.32, 0.55 \rangle & \langle 0.28, 0.55 \rangle & \langle 0.24, 0.70 \rangle & \langle \bot, \bot \rangle \\ k_2 & \langle 0.20, 0.70 \rangle & \langle 0.32, 0.55 \rangle & \langle 0.32, 0.60 \rangle & \langle \bot, \bot \rangle \\ k_3 & \langle 0.28, 0.65 \rangle & \langle 0.28, 0.60 \rangle & \langle 0.20, 0.65 \rangle & \langle \bot, \bot \rangle \\ Q & \langle \bot, \bot \rangle & \langle \bot, \bot \rangle & \langle \bot, \bot \rangle & \langle \bot, \bot \rangle \end{array} \right. ,$$

$$\begin{array}{c|cccc} h_2 & l_1 & l_2 & l_3 & R \\ \hline k_1 & \langle 0.21, 0.6 \rangle & \langle 0.15, 0.68 \rangle & \langle 0.12, 0.73 \rangle & \langle \bot, \bot \rangle \\ k_2 & \langle 0.18, 0.55 \rangle & \langle 0.21, 0.64 \rangle & \langle 0.10, 0.82 \rangle & \langle \bot, \bot \rangle \\ k_3 & \langle 0.21, 0.60 \rangle & \langle 0.15, 0.72 \rangle & \langle 0.12, 0.76 \rangle & \langle \bot, \bot \rangle \\ Q & \langle \bot, \bot \rangle & \langle \bot, \bot \rangle & \langle \bot, \bot \rangle & \langle \bot, \bot \rangle \end{array} ,$$

$$\begin{array}{c|cccc} h_3 & l_1 & l_2 & l_3 & R \\ \hline k_1 & \langle 0.24, 0.73 \rangle & \langle 0.18, 0.68 \rangle & \langle 0.07, 0.91 \rangle & \langle \bot, \bot \rangle \\ k_2 & \langle 0.21, 0.73 \rangle & \langle 0.18, 0.72 \rangle & \langle 0.06, 0.93 \rangle & \langle \bot, \bot \rangle \\ k_3 & \langle 0.18, 0.79 \rangle & \langle 0.15, 0.72 \rangle & \langle 0.07, 0.92 \rangle & \langle \bot, \bot \rangle \\ Q & \langle \bot, \bot \rangle & \langle \bot, \bot \rangle & \langle \bot, \bot \rangle & \langle \bot, \bot \rangle \end{array} ,$$

$$
\left.
\begin{array}{c|cccc}
T & l_1 & l_2 & l_3 & R \\
\hline
k_1 & \langle\perp,\perp\rangle & \langle\perp,\perp\rangle & \langle\perp,\perp\rangle & \langle\perp,\perp\rangle \\
k_2 & \langle\perp,\perp\rangle & \langle\perp,\perp\rangle & \langle\perp,\perp\rangle & \langle\perp,\perp\rangle \\
k_3 & \langle\perp,\perp\rangle & \langle\perp,\perp\rangle & \langle\perp,\perp\rangle & \langle\perp,\perp\rangle \\
Q & \langle\perp,\perp\rangle & \langle\perp,\perp\rangle & \langle\perp,\perp\rangle & \langle\perp,\perp\rangle
\end{array}
\right\} .
$$

Step 3. For each index k_i of K at a fixed h_g ($g = 1, 2, 3$) of the matrix S, the smallest element is found among the elements s_{k_i,l_j,h_g} ($j = 1, 2, 3$) and is recorded as the value of the element s_{k_i,R,h_g}. It is subtracted from all elements s_{k_i,l_j,h_g}, $j = 1, 2, 3$. We get the matrix S with reduced prices. Go to *Step 4*.

$$
S[K, L, H] = \left\{
\begin{array}{c|cccc}
h_1 & l_1 & l_2 & l_3 & R \\
\hline
k_1 & \langle 0.08, 0.92\rangle & \langle 0.04, 0.96\rangle & \langle 0.00, 1.00\rangle & \langle 0.24, 0.70\rangle \\
k_2 & \langle 0.00, 1,00\rangle & \langle 0.12, 0.88\rangle & \langle 0.12, 0.88\rangle & \langle 0.20, 0.70\rangle \\
k_3 & \langle 0.08, 0.92\rangle & \langle 0.08, 0.92\rangle & \langle 0.00, 1.00\rangle & \langle 0.20, 0.65\rangle \\
Q & \langle\perp,\perp\rangle & \langle\perp,\perp\rangle & \langle\perp,\perp\rangle & \langle\perp,\perp\rangle
\end{array}
\right. ,
$$

$$
\begin{array}{c|cccc}
h_2 & l_1 & l_2 & l_3 & R \\
\hline
k_1 & \langle 0.09, 0.91\rangle & \langle 0.03, 0.97\rangle & \langle 0.00, 1.00\rangle & \langle 0.12, 0.73\rangle \\
k_2 & \langle 0.08, 0.92\rangle & \langle 0.11, 0.89\rangle & \langle 0.00, 1.00\rangle & \langle 0.10, 0.82\rangle \\
k_3 & \langle 0.09, 0.91\rangle & \langle 0.03, 0.97\rangle & \langle 0.00, 1.00\rangle & \langle 0.12, 0.76\rangle \\
Q & \langle\perp,\perp\rangle & \langle\perp,\perp\rangle & \langle\perp,\perp\rangle & \langle\perp,\perp\rangle
\end{array} ,
$$

$$
\begin{array}{c|cccc}
h_3 & l_1 & l_2 & l_3 & R \\
\hline
k_1 & \langle 0.17, 0.83\rangle & \langle 0.11, 0.89\rangle & \langle 0.00, 1.00\rangle & \langle 0.07, 0.91\rangle \\
k_2 & \langle 0.15, 0.85\rangle & \langle 0.12, 0.88\rangle & \langle 0.00, 1.00\rangle & \langle 0.06, 0.93\rangle \\
k_3 & \langle 0.11, 0.89\rangle & \langle 0.08, 0.92\rangle & \langle 0.00, 1.00\rangle & \langle 0.07, 0.92\rangle \\
Q & \langle\perp,\perp\rangle & \langle\perp,\perp\rangle & \langle\perp,\perp\rangle & \langle\perp,\perp\rangle
\end{array} ,
$$

$$
\left.
\begin{array}{c|cccc}
T & l_1 & l_2 & l_3 & R \\
\hline
k_1 & \langle\perp,\perp\rangle & \langle\perp,\perp\rangle & \langle\perp,\perp\rangle & \langle\perp,\perp\rangle \\
k_2 & \langle\perp,\perp\rangle & \langle\perp,\perp\rangle & \langle\perp,\perp\rangle & \langle\perp,\perp\rangle \\
k_3 & \langle\perp,\perp\rangle & \langle\perp,\perp\rangle & \langle\perp,\perp\rangle & \langle\perp,\perp\rangle \\
Q & \langle\perp,\perp\rangle & \langle\perp,\perp\rangle & \langle\perp,\perp\rangle & \langle\perp,\perp\rangle
\end{array}
\right\} .
$$

Step 4. For each index l_j of L at a fixed h_g ($g = 1, 2, 3$) of the matrix S, the smallest element is found among the elements s_{k_i,l_j,h_g} ($i = 1, 2, 3$) and it is subtracted from all elements s_{k_i,l_j,h_g}, for $i = 1, 2, 3$. We get the matrix S with reduced costs and go to *Step 5*.

$$S[K, L, H] = \left\{ \begin{array}{l|cccc} h_1 & l_1 & l_2 & l_3 & R \\ \hline k_1 & \langle 0.08, 0.92 \rangle & \langle 0.00, 1.00 \rangle & \langle 0.00, 1.00 \rangle & \langle 0.24, 0.70 \rangle \\ k_2 & \langle 0.00, 1,00 \rangle & \langle 0.08, 0.92 \rangle & \langle 0.12, 0.88 \rangle & \langle 0.20, 0.70 \rangle \\ k_3 & \langle 0.08, 0.92 \rangle & \langle 0.04, 0.96 \rangle & \langle 0.00, 1.00 \rangle & \langle 0.20, 0.65 \rangle \\ Q & \langle 0.00, 1.00 \rangle & \langle 0.04, 0.96 \rangle & \langle 0.00, 1.00 \rangle & \langle \perp, \perp \rangle \end{array} \right. ,$$

$$\begin{array}{l|cccc} h_2 & l_1 & l_2 & l_3 & R \\ \hline k_1 & \langle 0.01, 0.99 \rangle & \langle 0.00, 1.00 \rangle & \langle 0.00, 1.00 \rangle & \langle 0.12, 0.73 \rangle \\ k_2 & \langle 0.00, 1.00 \rangle & \langle 0.08, 0.92 \rangle & \langle 0.00, 1.00 \rangle & \langle 0.10, 0.82 \rangle \\ k_3 & \langle 0.01, 0.99 \rangle & \langle 0.00, 1.00 \rangle & \langle 0.00, 1.00 \rangle & \langle 0.12, 0.76 \rangle \\ Q & \langle 0.08, 0.92 \rangle & \langle 0.03, 0.97 \rangle & \langle 0.00, 1.00 \rangle & \langle \perp, \perp \rangle \end{array} ,$$

$$\begin{array}{l|cccc} h_3 & l_1 & l_2 & l_3 & R \\ \hline k_1 & \langle 0.06, 0.94 \rangle & \langle 0.03, 0.97 \rangle & \langle 0.00, 1.00 \rangle & \langle 0.07, 0.91 \rangle \\ k_2 & \langle 0.04, 0.96 \rangle & \langle 0.04, 0.96 \rangle & \langle 0.00, 1.00 \rangle & \langle 0.06, 0.93 \rangle \\ k_3 & \langle 0.00, 1.00 \rangle & \langle 0.00, 1.00 \rangle & \langle 0.00, 1.00 \rangle & \langle 0.07, 0.92 \rangle \\ Q & \langle 0.11, 0.89 \rangle & \langle 0.08, 0.92 \rangle & \langle 0.00, 1.00 \rangle & \langle \perp, \perp \rangle \end{array} ,$$

$$\begin{array}{l|cccc} T & l_1 & l_2 & l_3 & R \\ \hline k_1 & \langle \perp, \perp \rangle & \langle \perp, \perp \rangle & \langle \perp, \perp \rangle & \langle \perp, \perp \rangle \\ k_2 & \langle \perp, \perp \rangle & \langle \perp, \perp \rangle & \langle \perp, \perp \rangle & \langle \perp, \perp \rangle \\ k_3 & \langle \perp, \perp \rangle & \langle \perp, \perp \rangle & \langle \perp, \perp \rangle & \langle \perp, \perp \rangle \\ Q & \langle \perp, \perp \rangle & \langle \perp, \perp \rangle & \langle \perp, \perp \rangle & \langle \perp, \perp \rangle \end{array} \Bigg\} .$$

Step 5. For each index k_i of K at a fixed l_j ($j = 1, 2, 3$) of the matrix S, the smallest element is found among the elements s_{k_i, l_j, h_g} ($g = 1, 2, 3$) and it is subtracted from all elements s_{k_i, l_j, h_g}, for $g = 1, 2, 3$ and go to *Step 6*.

$$S[K, L, H] = \left\{ \begin{array}{l|cccc} h_1 & l_1 & l_2 & l_3 & R \\ \hline k_1 & \langle 0.07, 0.93 \rangle & \langle 0.00, 1.00 \rangle & \langle 0.00, 1.00 \rangle & \langle 0.24, 0.70 \rangle \\ k_2 & \langle 0.00, 1,00 \rangle & \langle 0.04, 0.96 \rangle & \langle 0.12, 0.88 \rangle & \langle 0.20, 0.70 \rangle \\ k_3 & \langle 0.08, 0.92 \rangle & \langle 0.04, 0.96 \rangle & \langle 0.00, 1.00 \rangle & \langle 0.20, 0.65 \rangle \\ Q & \langle 0.00, 1.00 \rangle & \langle 0.04, 0.96 \rangle & \langle 0.00, 1.00 \rangle & \langle \perp, \perp \rangle \end{array} \right. ,$$

$$\begin{array}{l|cccc} h_2 & l_1 & l_2 & l_3 & R \\ \hline k_1 & \langle 0.00, 1.00 \rangle & \langle 0.00, 1.00 \rangle & \langle 0.00, 1.00 \rangle & \langle 0.12, 0.73 \rangle \\ k_2 & \langle 0.00, 1.00 \rangle & \langle 0.04, 0.96 \rangle & \langle 0.00, 1.00 \rangle & \langle 0.10, 0.82 \rangle \\ k_3 & \langle 0.01, 0.99 \rangle & \langle 0.00, 1.00 \rangle & \langle 0.00, 1.00 \rangle & \langle 0.12, 0.76 \rangle \\ Q & \langle 0.08, 0.92 \rangle & \langle 0.03, 0.97 \rangle & \langle 0.00, 1.00 \rangle & \langle \perp, \perp \rangle \end{array} ,$$

h_3	l_1	l_2	l_3	R
k_1	$\langle 0.05, 0.95 \rangle$	$\langle 0.03, 0.97 \rangle$	$\langle 0.00, 1.00 \rangle$	$\langle 0.07, 0.91 \rangle$
k_2	$\langle 0.04, 0.96 \rangle$	$\langle 0.00, 1.00 \rangle$	$\langle 0.00, 1.00 \rangle$	$\langle 0.06, 0.93 \rangle$,
k_3	$\langle 0.00, 1.00 \rangle$	$\langle 0.00, 1.00 \rangle$	$\langle 0.00, 1.00 \rangle$	$\langle 0.07, 0.92 \rangle$
Q	$\langle 0.11, 0.89 \rangle$	$\langle 0.08, 0.92 \rangle$	$\langle 0.00, 1.00 \rangle$	$\langle \perp, \perp \rangle$

T	l_1	l_2	l_3	R
k_1	$\langle 0.01, 0.99 \rangle$	$\langle 0.00, 1.00 \rangle$	$\langle 0.00, 1.00 \rangle$	$\langle \perp, \perp \rangle$
k_2	$\langle 0.00, 1.00 \rangle$	$\langle 0.04, 0.96 \rangle$	$\langle 0.00, 1.00 \rangle$	$\langle \perp, \perp \rangle$
k_3	$\langle 0.00, 1.00 \rangle$	$\langle 0.00, 1.00 \rangle$	$\langle 0.00, 1.00 \rangle$	$\langle \perp, \perp \rangle$
Q	$\langle \perp, \perp \rangle$	$\langle \perp, \perp \rangle$	$\langle \perp, \perp \rangle$	$\langle \perp, \perp \rangle$

Step 6. Cross out all elements $\langle 0, 1 \rangle$, in the matrix S with the minimum possible number of lines (horizontal, vertical or both).

$$S[K, L, H] =$$

h_1	l_1	l_2	l_3	R
k_1	$\langle 0.07, 0.93 \rangle$	$\langle 0.00, 1.00 \rangle$	$\langle 0.00, 1.00 \rangle$	$\langle 0.24, 0.70 \rangle$
k_2	$\langle 0.00, 1.00 \rangle$	$\langle 0.04, 0.96 \rangle$	$\langle 0.12, 0.88 \rangle$	$\langle 0.20, 0.70 \rangle$,
k_3	$\langle 0.08, 0.92 \rangle$	$\langle 0.04, 0.96 \rangle$	$\langle 0.00, 1.00 \rangle$	$\langle 0.20, 0.65 \rangle$
Q	$\langle 0.00, 1.00 \rangle$	$\langle 0.04, 0.96 \rangle$	$\langle 0.00, 1.00 \rangle$	$\langle \perp, \perp \rangle$

h_2	l_1	l_2	l_3	R
k_1	$\langle 0.00, 1.00 \rangle$	$\langle 0.00, 1.00 \rangle$	$\langle 0.00, 1.00 \rangle$	$\langle 0.12, 0.73 \rangle$
k_2	$\langle 0.00, 1.00 \rangle$	$\langle 0.04, 0.96 \rangle$	$\langle 0.00, 1.00 \rangle$	$\langle 0.10, 0.82 \rangle$,
k_3	$\langle 0.01, 0.99 \rangle$	$\langle 0.00, 1.00 \rangle$	$\langle 0.00, 1.00 \rangle$	$\langle 0.12, 0.76 \rangle$
Q	$\langle 0.08, 0.92 \rangle$	$\langle 0.03, 0.97 \rangle$	$\langle 0.00, 1.00 \rangle$	$\langle \perp, \perp \rangle$

h_3	l_1	l_2	l_3	R
k_1	$\langle 0.05, 0.95 \rangle$	$\langle 0.03, 0.97 \rangle$	$\langle 0.00, 1.00 \rangle$	$\langle 0.07, 0.91 \rangle$
k_2	$\langle 0.04, 0.96 \rangle$	$\langle 0.00, 1.00 \rangle$	$\langle 0.00, 1.00 \rangle$	$\langle 0.06, 0.93 \rangle$,
k_3	$\langle 0.00, 1.00 \rangle$	$\langle 0.00, 1.00 \rangle$	$\langle 0.00, 1.00 \rangle$	$\langle 0.07, 0.92 \rangle$
Q	$\langle 0.11, 0.89 \rangle$	$\langle 0.08, 0.92 \rangle$	$\langle 0.00, 1.00 \rangle$	$\langle \perp, \perp \rangle$

T	l_1	l_2	l_3	R
k_1	$\langle 0.01, 0.99 \rangle$	$\langle 0.00, 1.00 \rangle$	$\langle 0.00, 1.00 \rangle$	$\langle \perp, \perp \rangle$
k_2	$\langle 0.00, 1.00 \rangle$	$\langle 0.04, 0.96 \rangle$	$\langle 0.00, 1.00 \rangle$	$\langle \perp, \perp \rangle$
k_3	$\langle 0.00, 1.00 \rangle$	$\langle 0.00, 1.00 \rangle$	$\langle 0.00, 1.00 \rangle$	$\langle \perp, \perp \rangle$
Q	$\langle \perp, \perp \rangle$	$\langle \perp, \perp \rangle$	$\langle \perp, \perp \rangle$	$\langle \perp, \perp \rangle$

The test for the optimality is satisfied (the number of lines $= 9 = m.f = n.f$) and go to *Step 7*.

Step 7. An optimal solution has been found, in which assignments of the VMs to the tasks in a given time-moment are located where the elements $\langle 1, 0 \rangle$ in the X, with a single such element in each dimension.

Hence in a time moment:

- h_1: the cloudlet l_1 is allocated on VM k_2, l_2 on k_1 and, l_3 on k_3;
- h_2: the cloudlet l_1 is allocated on VM k_2, l_2 on k_1 and, l_3 on k_3;
- h_3: the cloudlet l_1 is allocated on VM k_3, l_2 on k_2 and, l_3 on k_1.

The optimal solution is in $X_{opt}[K*, L*, H * \{x_{k_i,l_j,h_g}\}]$:

$$
X_{opt}[K*, L*, H*] = \left\{ \begin{array}{c|ccc}
h_1 & l_1 & l_2 & l_3 \\
\hline
k_1 & \langle 0.00, 1.00 \rangle & \langle 1.00, 0.00 \rangle & \langle 0.00, 1.00 \rangle \\
k_2 & \langle 1.00, 0.00 \rangle & \langle 0.00, 1.00 \rangle & \langle 0.00, 1.00 \rangle \\
k_3 & \langle 0.00, 1.00 \rangle & \langle 0.00, 1.00 \rangle & \langle 1.00, 0.00 \rangle
\end{array} \right.,
$$

$$
\left. \begin{array}{c|ccc}
h_2 & l_1 & l_2 & l_3 \\
\hline
k_1 & \langle 0.00, 1.00 \rangle & \langle 1.00, 0.00 \rangle & \langle 0.00, 1.00 \rangle \\
k_2 & \langle 1.00, 0.00 \rangle & \langle 0.00, 1.00 \rangle & \langle 0.00, 1.00 \rangle \\
k_3 & \langle 0.00, 1.00 \rangle & \langle 0.00, 1.00 \rangle & \langle 1.00, 0.00 \rangle
\end{array} \right.
\quad
\begin{array}{c|ccc}
h_3 & l_1 & l_2 & l_3 \\
\hline
k_1 & \langle 0.00, 1.00 \rangle & \langle 0.00, 1.00 \rangle & \langle 1.00, 1.00 \rangle \\
k_2 & \langle 0.00, 1.00 \rangle & \langle 1.00, 0.00 \rangle & \langle 0.00, 1.00 \rangle \\
k_3 & \langle 1.00, 0.00 \rangle & \langle 0.00, 1.00 \rangle & \langle 0.00, 1.00 \rangle
\end{array} \left. \right\}.
$$

The optimal assignment cost is:

$$
AGIO_{\oplus_{(max,min)}} \left(C_{(\{Q\},\{R\},\{T\})} \otimes_{(min,max)} X_{opt} \right) = \langle 0.7, 0.1 \rangle.
$$

5 Conclusion

The main objective in cloud environment is to optimally allocate resources so as to achieve the performance of tasks with the lowest cost or time. IFIMs are an effective instrument for modeling the resource allocation process in cloud computing environment. In the paper we model the resource assignment process in uncertain cloud computing environment and propose the three-dimensional intuitinistic fuzzy Hungarian algorithm to find the optimal distribution of tasks to resources over time. The task priorities are defined in advance. This method solves 3D-IFAP directly without converting it into a classical linear problem. The time complexity of the proposed algorithm is comparable with that of the standard Hungarian algorithm. The proposed example demonstrates the correctness and efficiency of the algorithm. Its main advantages are that the algorithm can be applied to balanced and unbalanced problems with imprecise parameters and can be extended in order to obtain the optimal solution for other types of multidimensional assignment problems. In future the proposed method will be implemented for various types multidimensional optimization problems with fuzzy or intuitionistic fuzzy data.

References

1. Ahmad, A.: The best candidates method for solving optimization problems. J. Comput. Sci. **8**(5), 711–715 (2012)
2. Atanassov, K.T.: Intuitionistic Fuzzy Sets, VII ITKR Session, Sofia, 20-23 June 1983 (Deposed in Centr. Sci.-Techn. Library of the Bulg. Acad. of Sci. 1697/84) (in Bulgarian). Reprinted: Int. J. Bioautomation **20**(S1), S1–S6 (2016)
3. Atanassov, K.: Generalized index matrices. Comptes Rendus l'Academie Bulg. Sci. **40**(11), 15–18 (1987)
4. Atanassov, K.: On Generalized Nets Theory. "Prof. M. Drinov" Academic Publishing House, Sofia (2007)
5. Atanassov, K.: Temporal intuitionistic fuzzy graphs. Notes Intuit. Fuzzy Sets **4**(4), 59–61 (1998)
6. Atanassov, K.: Intuitionistic Fuzzy Sets. Springer, Heidelberg (1999)
7. Atanassov, K.: Generalized Nets. World Scientific, Singapore, London (1991)
8. Atanassov, K.: On Intuitionistic Fuzzy Sets Theory. STUDFUZZ, vol. 283. Springer, Heidelberg (2012). https://doi.org/10.1007/978-3-642-29127-2
9. Atanassov, K.: Index matrix representation of the intuitionistic fuzzy graphs. Fifth Scientific Session of the Mathematics Foundations of Artificial Intelligence Seminar, Sofia, Preprint MRL-MFAIS-10-94, pp. 36–41 (1994)
10. Atanassov, K.: On index matrix interpretations of intuitionistic fuzzy graphs. Notes Intuit. Fuzzy Sets **8**(4), 73–78 (2002)
11. Atanassov, K.: Index Matrices: Towards an Augmented Matrix Calculus. Studies in Computational Intelligence, vol. 573. Springer, Cham (2014). https://doi.org/10.1007/978-3-319-10945-9
12. Atanassov, K.: Intuitionistic Fuzzy Logics. Studies in Fuzziness and Soft Computing, vol. 351. Springer (2017). https://doi.org/10.1007/978-3-319-48953-7
13. Atanassov, K., Szmidt, E., Kacprzyk, J.: On intuitionistic fuzzy pairs. Notes Intuit. Fuzzy Sets **19**(3), 1–13 (2013)
14. Atanassov, K., Atanassova, V., Gluhchev, G.: InterCriteria analysis: ideas and problems. Notes Intuit. Fuzzy Sets **21**(1), 81–88 (2015)
15. Atanasova, V.: Marketing in tourism. Burgas **233** (2016)
16. Balas, E., Miller, D., Pekny, J., Toth, P.: A parallel shortest augmenting path algorithm for the assignment problem. J ACM (JACM) **38**(4), 985–1004 (1991)
17. Belacela, N., Boulasselb, M.: Multicriteria fuzzy assignment method: a useful tool to assist medical diagnosis. Artif. Intell. Med. **21**(1–3), 201–207 (2001)
18. Bertsekas, D.P., Castaon, D.A.: Parallel asynchronous Hungarian methods for the assignment problem. ORSA J Comput. **5**(3), 261–274 (1993)
19. Bilgaiyan, S., Sagnika, S., Das, M.: Workflow scheduling in cloud computing environment using cat swarm optimization. In: IEEE International Advance Computing Conference (IACC), pp. 680-685 (2014)
20. Bitam, S.: Bees life algorithm for job scheduling in cloud computing. In: International Conference on Computing and Information Technology (ICCIT) **186–191** (2012)
21. Bokhari, S.: A shortest tree algorithm for optimal assignments across space and time in a distributed processor system. IEEE Trans. Softw. Eng. **SE-7**(6) (1981)
22. Byzkan, G., Ger, F., Feyziolu, O.: Soft Computing **22**(15), 5091-5114 https://doi.org/10.1007/s00500-018-3317-4

23. Chaisiri, S., Lee, B., Niyato, D.: Optimization of resource provisioning cost in cloud computing. IEEE Trans. Serv. Comput. **5**(2), 164–176 (2012)
24. Feng, Y., Yang, L.: A two-objective fuzzy k-cardinality assignment problem. J. Comput. Appl. Math. **197**(1), 233–244 (2006)
25. Fidanova, S., Atanassova, V., Roeva, O.: Ant colony optimization application to GPS surveying problems: interCriteria analysis. In: Atanassov, K., et al. (eds.) Uncertainty and Imprecision in Decision Making and Decision Support: Cross-Fertilization, New Models and Applications. IWIFSGN 2016. Advances in Intelligent Systems and Computing, vol. 559, pp. 251–264. Springer, Cham (2018)
26. Fidanova, S., Paprzycki, M., Roeva, O.: Hybrid GA-ACO algorithm for a model parameters identification problem. In: Proceedings of the Federated Conference on Computer Science and Information Systems (FedCSIS), WCO, Poland, pp. 413–420 (2014). https://doi.org/10.15439/2014F373
27. Frieze, A.: Complexity of a 3-dimensional problem. Eur. J. Oper. Res. **13**(2), 161–164 (1983)
28. Huang, D., Chiu, H., Yeh, R., Chang, J.: A fuzzy multi-criteria decision making approach for solving a bi-objective personnel assignment problem. Comput. Ind. Eng. **56**(1), 1–10 (2009)
29. Huang, G., Lim, A.: A hybrid genetic algorithm for the three-index assignment problem. Eur. J. Oper. Res. **172**(1), 249–257 (2006)
30. Jonker, R., Volgenant, A.: Improving the Hungarian assignment algorithm. Oper. Res. Lett. **5**, 171–175 (1986)
31. Krishnadoss, P., Jacob, P.: OCSA: task scheduling algorithm in cloud computing environment. IInt. J. Intell. Eng. Syst. **11**(3), 273–279 (2018)
32. Kuhn, H.: On certain convex polyhedra. Bull. Am. Math. Soc. **61**, 557–558 (1955)
33. Kuhn, H.: The Hungarian method for the assignment problem. Nav. Res. Logist. Q. **2**, 83–97 (1955)
34. Kuhn, H.: The Hungarian method for the assignment problem. Nav. Res. Logist. Q. **3**, 253–258 (1956)
35. Kumar, N., Shukla, D.: Resource management through fuzzy assignment problem in cloud computing environment. In: Procceding ICTCS 16, Udaipur, India, 1–6 (2016). https://doi.org/10.1145/2905055.2905146
36. Lia, F., Xu, L., Jina, D., Wang, H.: Study on solution models and methods for the fuzzy assignment problems. Expert. Syst. Appl. **39**(12), 11276–11283 (2012)
37. Lin, C.: Assignment problem for team performance promotion under fuzzy environment. Math. Probl. Eng. 1–10 (2013)
38. Lin, C., Ma, K.: Model and algorithms of the fuzzy three-dimensional axial assignment problem with additional constraint. S. Afr. J. Ind. Eng. **26**(3), 54–70 (2015)
39. Little, J., Murty, K., Sweeney, D., Karel, C.: An algorithm for the travelling salesman problem. Oper. Res. **11**, 972–989 (1963)
40. Liu, F., Tong, J., Mao, J., Bohn, R., Messina, J., Badger, L., Leaf, D.: NIST Cloud Computing Reference Architecture. Recommendations of the National Institute of Standards and Technology, Special Publication, Gaithersburg, pp. 500–292 (2011)
41. Liu, L., Gao, X.: Fuzzy weighted equilibrium multi-job assignment problem and genetic algorithm. Appl. Math. Model. **33**(10), 3926–3935 (2009)
42. Liu, L., Li, Y.: The fuzzy quadratic assignment problem with penalty: new models and genetic algorithm. Appl. Math. Comput. **174**(2), 1229–1244 (2006)
43. Marinov, P., Fidanova, S.: Intercriteria and correlation Analyses: Similarities, Differences and Simultaneous Use. In: Annual of "Informatics" Section, Union of Scientists in Bulgaria (2015–2016)
44. Mukherjee, S., Basum, K.: Solution of a class of Intuitionistic fuzzy assignment problem by using similarity measures. Knowl. Based Syst. **27**, 170–179 (2012)
45. Munkres, J.: Algorithms for the assignment and transportation problems. J. Soc. Ind. Appl. Math. **5**(1), 32–38 (1957)
46. Nagarajan, R., Solairaju, A.: Computing improved fuzzy optimal Hungarian assignment problems with fuzzy costs under robust ranking techniques. Int. J. Comput. Appl. **6**(4), 6–13 (2010)

47. Olivo, P.: A mixed algorithm for the multidimensional assignment problem. Riv. Di Mat. Sci. Econ. Sociali **6**, 67–78 (1983)
48. Pandey, S., Wu, L., Guru, M., Buyya, R.: A particle swarm optimization-based heuristic for scheduling workflow applications in cloud computing environments. In: 24th IEEE International Conference on Advanced Information Networking and Applications, pp. 400–407 (2010)
49. Pandian, P., Kavitha, K.: A new method for solving fuzzy assignment problem. Ann. Pure Appl. Math. **1**(1), 69–83 (2012)
50. Paul, M., Sanyal, G.: Task scheduling in cloud computing using credit based assignment problem. Int. J. Comput. Sci. Eng. **3**(10), 3426–3430 (2011)
51. Pierskalla, W.: The multidimensional assignment problem. Oper. Res. **16**, 422–430 (1968)
52. Prabakaran, K., Ganesan, K.: Fuzzy Hungarian method for solving intuitionistic fuzzy assignment problems. Int. J. Sci. Eng. Res. **6**(3), 11–17 (2015)
53. Prabha, S., Vimala, S.: Optimal solution for the intuitinstic fuzzy assignment problem via three methods—IFRMM. IFOAM. IFAM. Adv. Res. **7**(6), 1–8 (2016)
54. Rajaraman, K., Sophia Porchelvi, R., Irene Hepzibah, R.: Multicriteria decision making in marketing mix on customer satisfaction using triangular intuitionistic fuzzy numbers. Int. J. Pure Appl. Math. **6**, 371–379 (2018)
55. Rathi, K., Balamohan, R., Shanmugasundaram, P., Revathi, M.: Fuzzy row penalty method to solve assignment problems with uncertain parameters. Glob. J. Pure Appl. Math. **11**(1), 39–44 (2015)
56. Ribagin, S., Atanassov, K., Roeva, O., Pencheva, T.: Generalized net model of adolescent idiopathic scoliosis diagnosing, uncertainty and imprecision in decision making and decision support: cross-fertilization, new models and applications (Atanassov K. T., J. Kacprzyk, A. Kauszko, M. Krawczak, J. Owsiski, S. Sotirov, E. Sotirova, E. Szmidt, S. Zadrony, Eds). Adv. Intell. Syst. Comput. **559**, 333–348 (2018)
57. Roeva, O., Vassilev, P., Ikonomov, N., Angelova, M., Su, J., Pencheva, T.: On different algorithms for interCriteria relations calculation. In: Hadjiski M., Atanassov, K.T. (eds.) Intuitionistic Fuzziness and Other Intelligent Theories and Their Applications. Studies in Computational Intelligence, vol. 757, pp. 143–160 (2019)
58. Roeva, O., Fidanova, O., Vassilev, P., Gepner, P.: InterCriteria analysis of a model parameters identification using genetic algorithm. In: Proceedings of the Federated Conference on Computer Science and Information Systems. Annals of Computer Science and Information Systems, vol. 5, pp. 501–506 (2015). https://doi.org/10.15439/2015F223
59. Sagnika, S., Bilgaiyan, S., Mishra, B. Workflow scheduling in cloud computing environment using bat algorithm. In: Somani A., Srivastava S., Mundra A., Rawat S. (eds.) Proceedings of First International Conference on Smart System, Innovations and Computing. Smart Innovation, Systems and Technologies, vol. 79. Springer, Singapore, pp. 149–163 (2018)
60. Shakya, K., Karaulia, D.: Survey on virtual machine scheduling in cloud environment. Int. J. Adv. Res. Comput. Sci. Softw. Eng. **4**(2) (2014)
61. Shrinivasan, A.: Method for solving fuzzy assignment problem using ones assignment method and robusts ranking technique. J. Appl. Math. Sci. **7**(113), 5607–5619 (2013)
62. Simeonov, S., Atanassova, V., Sotirova, E., Simeonova, N., Kostadinov, T.: Generalized net of a centralized embedded system. Uncertainty and imprecision in decision making and decision support: cross-fertilization, new models and applications. IWIFSGN 2016. Advances in Intelligent Systems and Computing, vol. 559, pp. 299–304. Springer, Cham (2018)
63. Singh, A., Juneja, D., Malhotra, M.: A novel agent based autonomous and service composition framework for cost optimization of resource provisioning in cloud computing. J. King Saud Univ. Comput. Inf. Sci. **29**(1), 19–28 (2015)
64. Sotirov, S., Sotirova, E., Werner, M., Simeonov, S., Hardt, W., Simeonova, N.: Ituitionistic Fuzzy Estimation of the Generalized Nets Model of Spatial-Temporal Group Scheduling Problems. Imprecision Uncertain. Inf. Represent. Process. Ser. Stud. Fuzziness Soft Comput. **332**, 401–414 (2016)
65. Sotirov, S., Werner, M., Simeonov, S., Hardt, W., Sotirova, E., Simeonova, N.: Using Generalized nets to Model Spatial-temporal Group Scheduling Problems. Issues IFSs GNs SRI-PAS **11**, 42–54 (2014)

66. Sotirov, S., Sotirova, E., Atanassova, V., Atanassov, K., et al.: A hybrid approach for modular neural network design using intercriteria analysis and intuitionistic fuzzy logic. Complexity. **2018** (2018). https://doi.org/10.1155/2018/3927951
67. Stone, H.: Multiprocessor scheduling with the aid of network flow algorithms. IEEE Trans. Softw. Eng. **SE-3**(1) (1977)
68. SundarRajan, R., Vasudevan, V., Mithya, S.: Workflow scheduling in cloud computing environment using firefly algorithm. In: Proceedings of International Conference on Electrical, Electronics, and Optimization Techniques (ICEEOT), pp. 955–960 (2016)
69. Traneva, V.: On 3-dimensional intuitionistic fuzzy index matrices. Notes Intuit. Fuzzy Sets **20**(4), 59–64 (2014)
70. Traneva, V.: Internal operations over 3-dimensional extended index matrices. In: Proceedings of the Jangjeon Mathematical Society, vol. 18, no. 4, pp. 547–569 (2015)
71. Traneva, V., Tranev, S.: Index Matrices as a Tool for Managerial Decision Making. Publishing House of the Union of Scientists, Bulgaria (2017) (in Bulgarian)
72. Traneva, V., Tranev, S., Atanassova, V.: An Intuitionistic Fuzzy Approach to the Hungarian Algorithm. Springer Nature Switzerland AG. Nikolov, G., et al. (eds.): NMA 2018, LNCS 11189, 19 (2019) https://doi.org/10.1007/978-3-030-10692-8_19
73. Wright, M.: Speeding up the Hungarian algorithm. Comput. Oper. Res. **17**(1), 95–96 (1990)
74. Zadeh, L.: Fuzzy sets. Inf. Control. **8**(3), 338–353 (1965)
75. Zoteva, D., Atanassova, V., Roeva, O., Szmidt, E.: Generalized net model of artificial bee colony optimization algorithm. In: Proceedings of ANNA'18; Advances in Neural Networks and Applications (2018)
76. Software for index matrices. http://justmathbg.info/indmatcalc.html. Accessed 1 Feb 2019
77. Munkres' Assignment Algorithm. http://csclab.murraystate.edu/bob.pilgrim/445//munkres.html. Accessed 8 May 2018

Three-Dimensional Interval-Valued Intuitionistic Fuzzy Appointment Model

Velichka Traneva, Vassia Atanassova and Stoyan Tranev

Abstract The paper explores the process of appointment of positions in an organization in conditions of uncertainty. We extend the optimization problem [28] of the process of appointment and reappointment, based on partial knowledge about the values of evaluation criteria of the human resources over time. Here, the 3-dimensional optimal appointment problem is formulated and an algorithm for its optimal solution is proposed, where the evaluations of candidates against criteria formulated by several experts at a fixed time, are interval-valued intuitionistic fuzzy pairs (IVIFPs). The proposed algorithm for the solution takes into account the ratings of the experts and the weight coefficients of the assessment criteria according to their priority for the respective position.

1 Introduction

Selection, appointment and evaluation of staff in an organization is a key moment in management [22]. Modern environment is uncertain and the parameters involved in the selection of human resources are fuzzy. To overcome this, Zadeh [29] introduced fuzzy set concept to deal with imprecision and vagueness. Thus, fixed values of membership and non-membership cannot handle such uncertainty involved in real-life problems. Atanassov and Gargov in 1989 [11] first proposed the concept of interval-valued intuitionist fuzzy sets (IVIFSs), which are an extension of both

V. Traneva (✉) · S. Tranev
"Prof. Asen Zlatarov" University, "Prof. Yakimov" Blvd, 8000 Bourgas, Bulgaria
e-mail: veleka13@gmail.com

S. Tranev
e-mail: tranev@abv.bg

V. Atanassova
Institute of Biophysics and Biomedical Engineering, Bulgarian Academy of Sciences,
105 "Acad. G. Bonchev" Str., 1113 Sofia, Bulgaria
e-mail: vassia.atanassova@gmail.com

S. Fidanova (ed.), *Recent Advances in Computational Optimization*,
Studies in Computational Intelligence 838,
https://doi.org/10.1007/978-3-030-22723-4_12

the intuitionistic fuzzy sets (IFSs) [7] and interval-valued fuzzy sets [21]. It was introduced in [2, 11] and described in detail in [4, 5, 10].

Two-dimensional problems of appointment to vacant positions in an organization by IFSs have been investigated in [18, 20], and in [16]. In [18] a new area of application of intuitionistic fuzzy sets theory was proposed in appointment of positions in an organisation using intuitionistic fuzzy sets approach via max-min-max rule. In [20] the concept of intuitionistic fuzzy multisets was applied to appointment procedure or process using a new distance measure in [19]. In [16] a model was presented, which describes a method for solving the problem of appointments using the theories of index matrices [1] and generalized nets [3, 6]. The three-dimensional optimal appointment problem was formulated in [28] and an algorithm for its optimal solution was proposed, where the evaluations of candidates against criteria formulated by several experts in a certain time (or location), are intuitionistic fuzzy pairs.

This paper, which is an extension of [28] explores for the first time the three-dimensional model in time during of the process of evaluation and selection of staff for vacancies in the organization, using the concepts of index matrices (IMs) [13] and IVIFSs. The evaluation is done by experts. The rating of each expert is assigned and is in the form of IVIFP [12], and the candidates' estimates on the criteria for the job are IVIFPs.

The rest of the paper is structured as follows. In Sect. 2, we present shortly the concepts of IMs and of IVIFPs. In Sect. 3, we define the three-dimensional interval-valued intuitionistic fuzzy appointment problem (3D-IVIFAP), and propose an algorithm for its optimal solution, in which evaluations of candidates against the criteria set by experts in a certain moment time are IVIFPs. The proposed example in Sect. 4 demonstrates the correctness and efficiency of the algorithm. Section 5 offers the conclusions and outlines some aspects for future research.

2 Basic Definitions

This section provides some definitions on IVIFPs (see [7, 10, 12]) and on index matrices (see [8, 9, 24]).

2.1 Short Notes on Interval-Valued Intuitionistic Fuzzy Pairs

The concept of IVIFPs was introduced in [12]. The **IVIFP** is an object of the form $\langle M, N \rangle$, where $M, N \subseteq [0, 1]$ are closed sets, $M = \left[\inf M, \sup M\right], N = \left[\inf N, \sup N\right]$ and

$$\sup M + \sup N \leq 1,$$

that is used as an evaluation of some object or process and whose components (M and N) are interpreted as intervals of degrees of membership and non-membership, or intervals of degrees of validity and non-validity, or intervals of degree of correctness and non-correctness, etc.

The IVIFP x is an **"interval-valued intuitionistic fuzzy false pair"** (**IVIFFP**) if and only if inf $M \leq$ sup N, while x is a **"false pair"** (**IVFP**) iff $M = [0, 0]$, $N = [1, 1]$.

The IVIFP x is an **"interval-valued intuitionistic fuzzy tautological pair"** (**IVIFTP**) if and only if inf $M \geq$ sup N, while x is a **"tautological pair"** (**IVTP**) iff $M = [1, 1]$, $N = [0, 0]$.

Let us have two IVIFPs $x = \langle M, N \rangle$ and $y = \langle P, Q \rangle$. In [10, 12] are defined the operations classical negation, conjunction, disjunction, multiplication with constant, and difference

$$\neg x = \langle N, M \rangle;$$
$$x \wedge_1 y = \qquad \langle [\min(\inf M, \inf P), \min(\sup M, \sup P)],$$
$$[\max(\inf N, \inf Q), \max(\sup N, \sup Q)]\rangle;$$
$$x \vee_1 y \qquad = \langle [\max(\inf M, \inf P), \max(\sup M, \sup P)],$$
$$[\min(\inf N, \inf Q), \min(\sup N, \sup Q)]\rangle;$$
$$x \wedge_2 y = x + y = \qquad \langle [\inf M + \inf P - \inf M \inf P,$$
$$\sup M + \sup P - \sup M \sup P], [\inf N \inf Q, \sup N \sup Q]\rangle;$$
$$x \vee_2 y = x.y = \qquad \langle [\inf M \inf P, \sup M \sup P], [\inf N + \inf Q - \inf N \inf Q,$$
$$\sup N + \sup Q - \sup N \sup Q]\rangle;$$
$$x \vee_3 y = \qquad \langle \left[\tfrac{\inf M + \inf P}{2}, \tfrac{\sup M + \sup P}{2} \right],$$
$$\langle \left[\tfrac{\inf P + \inf Q}{2}, \tfrac{\sup P + \sup Q}{2} \right] \rangle.$$
$$\alpha.x = \qquad \langle [1 - (1 - \inf M)^\alpha, 1 - (1 - \sup,)^\alpha], [\inf N^\alpha, \sup N^\alpha] \rangle \, (\alpha \in R)$$
$$x - y = \qquad \langle [\max(0, \min(\inf M - \sup P, 1 - \sup N + \sup Q)),$$
$$\max(0, \min(\sup M - \inf P, 1 - \sup N + \sup Q))],$$
$$[\min(1, \min(\inf N + \inf Q, 1 - \sup M + \inf P)),$$
$$\min(1, \min(\sup N + \sup Q, 1 - \sup M + \inf P))]\rangle$$

$$(1)$$

and the relations

$$
\begin{array}{lll}
x \geq y & \text{iff} & \inf M \geq \inf P \text{ and } \sup M \geq \sup P \\
 & & \text{and } \inf N \leq \inf Q \text{ and } \sup N \leq \sup Q \\
x \geq_\square y & \text{iff} & \inf M \geq \inf P \text{ and } \sup M \geq \sup P \\
x \geq_\diamond y & \text{iff} & \inf N \leq \inf Q \text{ and } \sup N \leq \sup Q \\[4pt]
x \leq y & \text{iff} & \inf M \leq \inf P \text{ and } \sup M \leq \sup P \\
 & & \text{and } \inf N \geq \inf Q \text{ and } \sup N \geq \sup Q \\
x \leq_\square y & \text{iff} & \inf M \leq \inf P \text{ and } \sup M \leq \sup P \\
x \leq_\diamond y & \text{iff} & \inf N \geq \inf Q \text{ and } \sup N \geq \sup Q \\[4pt]
x = y & \text{iff} & \inf M = \inf P \text{ and } \sup M = \sup P \\
 & & \text{and } \inf Q = \inf N \text{ and } \sup Q = \sup N \\
x =_\square y & \text{iff} & \inf M = \inf P \text{ and } \sup M = \sup P \\
x =_\diamond y & \text{iff} & \inf N = \inf Q \text{ and } \sup N = \sup Q.
\end{array}
$$

$$(2)$$

Let a set E be fixed. An **"interval-valued intuitionistic fuzzy set"** (IVIFS) A in E is an object of the following form (see [5, 11]):

$$A = \{\langle x, M_A(x), N_A(x)\rangle | x \in E\},$$

where

$$M_A \subseteq [0, 1] \text{ and } N_A \subseteq [0, 1]$$

are intervals and for all $x \in E$:

$$0 \leq \sup M_A(x) + \sup N_A(x) \leq 1.$$

2.2 Definition of 3D-IVIFIM, Operations and Relations

The concept of index matrices (IMs) was introduced in 1984 in [1]. In this section, following [8, 9, 24], we introduce three-dimensional interval-valued intuitionistic fuzzy index matrix (3D-IVIFIM). Let $\mathscr{I}$ be a fixed set. By 3D-IVIFIM with index sets K, L and H $(K, L, H \subset \mathscr{I})$, we denote the object:

$$[K, L, H, \{\langle M_{k_i,l_j,h_g}, N_{k_i,l_j,h_g}\rangle\}]$$

$$\equiv \begin{array}{c|cccccc}
h_g \in H & l_1 & \cdots & l_j & \cdots & l_n \\
\hline
k_1 & \langle M_{k_1,l_1,h_g}, M_{k_1,l_1,h_g}\rangle & \cdots & \langle M_{k_1,l_j,h_g}, N_{k_1,l_j,h_g}\rangle & \cdots & \langle M_{k_1,l_n,h_g}, N_{k_1,l_n,h_g}\rangle \\
\vdots & \vdots & \ddots & \vdots & \ddots & \vdots \\
k_m & \langle M_{k_m,l_1,h_g}, N_{k_m,l_1,h_g}\rangle & \cdots & \langle M_{k_m,l_j,h_g}, N_{k_m,l_j,h_g}\rangle & \cdots & \langle M_{k_m,l_n,h_g}, N_{k_m,l_n,h_g}\rangle
\end{array},$$

where for every $1 \leq i \leq m, 1 \leq j \leq n, 1 \leq g \leq f$:

$$M_{k_i,l_j,h_g} \subseteq [0, 1], N_{k_i,l_j,h_g} \subseteq [0, 1], \sup M_{k_i,l_j,h_g} + \sup N_{k_i,l_j,h_g} \leq 1.$$

Let "$*$" and "$\circ$" be two fixed operations over IVIFPs and let $\langle M, N\rangle * \langle P, Q\rangle = \langle M *_l P, N *_r Q\rangle$, where "$*_l$" and "$*_r$" are determined by the form of the operation "$*$". Following [8, 9, 24], we extend some operations over two 3D-IVIFIMs $A = [K, L, H, \{\langle M_{k_i,l_j,h_g}, N_{k_i,l_j,h_g}\rangle\}]$ and $B = [P, Q, E, \{\langle R_{p_r,q_s,e_d}, S_{p_r,q_s,e_d}\rangle\}]$:
Negation:

$$\neg A = [K, L, H, \{\langle N_{k_i,l_j,h_g}, M_{k_i,l_j,h_g}\rangle\}].$$

Addition-(*):

$$A \oplus_{(*_l,*_r)} B = [K \cup P, L \cup Q, H \cup E, \{\langle \Phi_{t_u,v_w,x_y}, \Psi_{t_u,v_w,x_y}\rangle\}],$$

where

$$\langle \Phi_{t_u,v_w,x_y}, \Psi_{t_u,v_w,x_y} \rangle$$

$$= \begin{cases} \langle M_{k_i,l_j,h_g}, N_{k_i,l_j,h_g} \rangle, & \text{if } t_u = k_i \in K, v_w = l_j \in L - Q, x_y = h_g \in H - E \\ & \text{or } t_u = k_i \in K, v_w = l_j \in L - Q, x_y = h_g \in H \\ & \text{or } t_u = k_i \in K - P, v_w = l_j \in L, x_y = h_g \in H \\[2mm] \langle R_{p_r,q_s,e_d}, S_{p_r,q_s,e_d} \rangle, & \text{if } t_u = p_r \in P, v_w = q_s \in Q, x_y = e_d \in E - H \\ & \text{or } t_u = p_r \in P, v_w = q_s \in Q - L, x_y = e_d \in E \\ & \text{or } t_u = p_r \in P - K, v_w = q_s \in Q, x_y = e_d \in E \\[2mm] \langle *_l(M_{k_i,l_j,h_g}, R_{p_r,q_s,e_d}), & \text{if } t_u = k_i = p_r \in K \cap P, v_w = l_j = q_s \in L \cap Q \\ \quad *_r(N_{k_i,l_j,h_g}, S_{p_r,q_s,e_d}) \rangle, & \text{and } x_y = h_g = e_d \in H \cap E \\[2mm] \langle [0, 0], [1, 1] \rangle, & \text{otherwise} \end{cases}$$

Termwise multiplication-$(*)$**:**

$$A \otimes_{(*_l,*_r)} B = [K \cap P, L \cap Q, H \cap R, \{\langle \Phi_{t_u,v_w,x_y}, \Psi_{t_u,v_w,x_y} \rangle\}],$$

where

$$\langle \Phi_{t_u,v_w,x_y}, \Psi_{t_u,v_w,x_y} \rangle = \langle *_l(M_{k_i,l_j,h_g}, R_{p_r,q_s,e_d}), *_r(N_{k_i,l_j,h_g}, S_{p_r,q_s,e_d}) \rangle.$$

Reduction: We use symbol "$\perp$" for lack of some component in the separate definitions. In some cases, it is suitable to change this symbol with "0". The operation $(k, \perp, \perp)$-reduction of a given IM A is defined by:

$$A_{(k,\perp,\perp)} = [K - \{k\}, L, H, \langle \Phi_{t_u,v_w,x_y}, \Psi_{t_u,v_w,x_y} \rangle],$$

where

$$\langle \Phi_{t_u,v_w,x_y}, \Psi_{t_u,v_w,x_y} \rangle = \langle M_{k_i,l_j,h_g}, N_{k_i,l_j,h_g} \rangle$$

for

$$t_u = k_i \in K - \{k\}, v_w = l_j \in L \text{ and } x_y = h_g \in H.$$

The definitions of the $(\perp, l, \perp)$- and $(\perp, \perp, h)$-reductions are analogous.
Multiplication with a constant α:

$$\alpha A = [K, L, H\{\alpha \langle M_{k_i,l_j,h_g}, N_{k_i,l_j,h_g} \rangle\}],$$

If α is a real number, then following [8, 17]

$$\alpha \langle M_{k_i,l_j,h_g}, N_{k_i,l_j,h_g} \rangle$$

$$= \left\langle [1 - (1 - \inf M_{k_i,l_j,h_g})^\alpha, 1 - (1 - \sup M_{k_i,l_j,h_g})^\alpha], [\inf N^\alpha_{k_i,l_j,h_g}, \sup N^\alpha_{k_i,l_j,h_g}] \right\rangle.$$

If $\alpha = \langle A, B \rangle$ is an IVIFP, then

$$\alpha \langle M_{k_i,l_j,h_g}, N_{k_i,l_j,h_g} \rangle = \left\langle [\inf A \inf M_{k_i,l_j,h_g}, \sup A \sup M_{k_i,l_j,h_g}], \right.$$

$$\left. [\inf B + \inf N_{k_i,l_j,h_g} - \inf B \inf N_{k_i,l_j,h_g}, \sup B + \sup N_{k_i,l_j,h_g} - \sup B \sup N_{k_i,l_j,h_g}] \right\rangle.$$

Multiplication-$(\circ, *)$:

$$A \odot_{(\circ,*)} B = [K \cup (P - L), Q \cup (L - P), H \cup R, \{\langle \Phi_{t_u,v_w,x_y}, \Psi_{t_u,v_w,x_y} \rangle\}],$$

where

$$\langle \Phi_{t_u,v_w,x_y}, \Psi_{t_u,v_w,x_y} \rangle =$$

$$= \begin{cases} \langle M_{k_i,l_j,h_g}, N_{k_i,l_j,h_g} \rangle, & \begin{aligned} &\text{if } t_u = k_i \in K \\ &\& v_w = l_j \in L - P - Q \ \& \ x_y = h_g \in H \\ &\text{or } t_u = k_i \in K - P - Q \\ &\& v_w = l_j \in L \ \& \ x_y = h_g \in H; \end{aligned} \\[2ex] \langle R_{p_r,q_s,r_d}, S_{p_r,q_s,r_d} \rangle, & \begin{aligned} &\text{if } t_u = p_r \in P \\ &\& v_w = q_s \in Q - K - L \ \& \ x_y = r_d \in R \\ &\text{or } t_u = p_r \in P - L - K \\ &\& v_w = q_s \in Q \ \& \ x_y = r_d \in R; \end{aligned} \\[2ex] \left\langle \underset{l_j=p_r\in L\cap P}{\circ_l} (*_l(M_{k_i,l_j,h_g}, R_{p_r,q_s,r_d})), \right. & \text{if } t_u = k_i \in K \ \& \ v_w = q_s \in Q \\[2ex] \left. \underset{l_j=p_r\in L\cap P}{\circ_r} (*_r(N_{k_i,l_j,h_g}, S_{p_r,q_s,r_d})) \right\rangle, & \& \ x_y = h_g = r_d \in H \cap R; \\[1ex] \langle [0, 0], [1, 1] \rangle, & \text{otherwise.} \end{cases}$$

where "$*$" and "$\circ$" are two fixed operations from (1) over IVIFPs.

Projection: Let A be an 3D-IVIFIM and $W \subseteq K$, $V \subseteq L$ and $U \subseteq H$ and. Then,

$$pr_{W,V,U}A = [W, V, U, \{\langle R_{p_r,q_s,e_d}, S_{p_r,q_s,e_d} \rangle\}],$$

where for each $k_i \in W$, $l_j \in V$ and $h_g \in U$,

$$\langle R_{p_r,q_s,e_d}, S_{p_r,q_s,e_d} \rangle = \langle M_{k_i,l_j,h_g}, N_{k_i,l_j,h_g} \rangle.$$

Substitution: Let IM $A = [K, L, H, \{a_{k,l,h}\}]$ be given. Local substitution over A is defined for the couple of indices (p, k) by

$$\left[\frac{p}{k}; \perp; \perp \right] A = \left[(K - \{k\}) \cup \{p\}, L, H, \{a_{k,l,h}\} \right].$$

The definitions of operations $\left[\bot; \frac{q}{l}; \bot\right] A$ and $\left[\bot; \bot; \frac{e}{h}\right] A$, where (q, l) and (e, h) are couples of indices, are analogous.

Aggregation operation by one dimension: Let us have two IVIFPs $x = \langle A, B \rangle$ and $y = \langle C, D \rangle$.

We use following three operations $\#_q$, $(q \le i \le 3)$ for scaling aggregation evaluations:

$$x\#_1 y = \langle \left[min(\inf A, \inf C), min(\sup A, \sup C)\right],$$

$$\left[max(\inf B, \inf D), max(\sup B, \sup D)\right]\rangle;$$

$$x\#_2 y = \langle \left[average(\inf A, \inf C), average(\sup A, \sup C)\right],$$

$$\left[average(\inf B, \inf D), average(\sup B, \sup D)\right]\rangle;$$

$$x\#_3 y = \langle \left[max(\inf A, \inf C), max(\sup A, \sup C)\right],$$

$$\left[min(\inf B, \inf D), min(\sup B, \sup D)\right]\rangle;$$

From the theorem in [27] we can see that: $x\#_1 y \le x\#_2 y \le x\#_3 y$.

Let $h_0 \notin H$ be a fixed index and $A = [K, L, H, \{\langle M_{k_i,l_j,h_g}, N_{k_i,l_j,h_g}\rangle\}]$ is an IVIFIM.

Following [9, 26, 27], we extend the definition of the aggregation operation by the dimension H:

$$\alpha_{H,\#_q}(A, h_0) = \begin{array}{c|ccc} h_0 & l_1 & \cdots & l_n \\ \hline k_1 & \overset{f}{\underset{g=1}{\#_q}} \langle M_{k_1,l_1,h_g}, N_{k_1,l_1,h_g}\rangle & \cdots & \overset{f}{\underset{g=1}{\#_q}} \langle M_{k_1,l_n,h_g}, N_{k_1,l_n,h_g}\rangle \\ \vdots & \vdots & \ddots & \vdots \\ k_i & \overset{f}{\underset{g=1}{\#_q}} \langle M_{k_i,l_1,h_g}, N_{k_i,l_1,h_g}\rangle & \cdots & \overset{f}{\underset{g=1}{\#_q}} \langle M_{k_i,l_n,h_g}, N_{k_i,l_n,h_g}\rangle \\ \vdots & \vdots & \ddots & \vdots \\ k_m & \overset{f}{\underset{g=1}{\#_q}} \langle M_{k_m,l_1,h_g}, N_{k_m,l_1,h_g}\rangle & \cdots & \overset{f}{\underset{g=1}{\#_q}} \langle M_{k_m,l_n,h_g}, N_{k_m,l_n,h_g}\rangle \end{array}, \quad (3)$$

where $1 \le q \le 3$.

The non-strict relation "inclusion about value" is defined by:

$$A \subseteq_v B \text{ iff } (K = P)\&(L = Q)\&(H = R)\&(\forall k \in K)(\forall l \in L)(\forall h \in H)$$

$$(\langle M_{k,l,h}, N_{k,l,h}\rangle \le \langle R_{k,l,h}, S_{k,l,h}\rangle).$$

3 Three-Dimensional Interval-Valued Intuitionistic Fuzzy Appointment Model

In the section, we will extend the 3D-intuitionistic fuzzy appointment problem from [28] by the IVIFSs and IVIFIMs concepts, and will propose an algorithm for its optimal solution.

3.1 Problem Formulation

An organization has u vacant positions $\{v_1, \ldots, v_e, \ldots, v_u\}$. Employees who have never worked in the organization apply to these positions, but there are those who are currently working or have ever worked in it. A system for evaluating the staff by criteria $\{c_1, \ldots, c_j, \ldots, c_n\}$ in a time-moment h_g (for $1 \leq g \leq f$) by the experts $\{d_1, \ldots, d_s, \ldots, d_D\}$ operates in the organization. The candidates $\{k_1, \ldots, k_i, \ldots, k_m\}$ for the positions occupying or having similar positions in the organization have been periodically evaluated according to the criteria for the fulfillment of their official duties and their estimates es_{k_i,c_j,d_s} (for $1 \leq i \leq m, 1 \leq j \leq n, 1 \leq s \leq D$) are IVIFPs. At the moment of applying for vacancies, all candidates are evaluated by the experts according to the criteria for job hiring, and their evaluations es_{k_i,c_j,d_s} (for $1 \leq i \leq m, 1 \leq j \leq n, 1 \leq s \leq D$) are IVIFPs. The ratings of the experts $\{r_1, \ldots, r_w, \ldots, r_D\}$ given (r_w for $1 \leq w \leq D$ is IVIFP), as well as the weight coefficients of the assessment criteria c_j (for $1 \leq j \leq n$) according to their priority for the respective position v_e (for $1 \leq e \leq u$)—pk_{c_j,v_e} and they are also IVIFPs. The aim of the problem is to optimally allocate vacancies among the candidates.

3.2 An Algorithm for Finding an Optimal Solution to the Problem

Let us create the following 3D-IVIFIM, in accordance with the problem in Sect. 3.1:

$$ES[K, C, D, \{es_{k_i,c_j,d_s}\}]$$

$$= \begin{array}{c|cccccc}
d_s \in D & c_1 & \cdots & c_j & \cdots & c_n \\
\hline
k_1 & \langle M_{k_1,c_1,d_s}, N_{k_1,c_1,d_s}\rangle & \cdots & \langle M_{k_1,c_j,d_s}, N_{k_1,c_j,d_s}\rangle & \cdots & \langle M_{k_1,c_n,d_s}, N_{k_1,c_n,d_s}\rangle \\
\vdots & \vdots & \ddots & \vdots & \ddots & \vdots \\
k_m & \langle M_{k_m,c_1,d_s}, N_{k_m,c_1,d_s}\rangle & \cdots & \langle M_{k_m,c_j,d_s}, N_{k_m,c_j,d_s}\rangle & \cdots & \langle M_{k_m,c_n,d_s}, N_{k_m,c_n,d_s}\rangle
\end{array},$$

where $K = \{k_1, k_2, \ldots, k_m\}$, $C = \{c_1, c_2, \ldots, c_n\}$, $D = \{d_1, d_2, \ldots, d_D\}$ and the element $\{es_{k_i,c_j,d_s}\} = \langle M_{k_i,c_j,d_s}, N_{k_i,c_j,d_s}\rangle$ (for $1 \leq i \leq m, 1 \leq j \leq n, 1 \leq s \leq D$) is the estimate of the d_s-th expert for the k_i-th employee by the c_j-th criterion and is an IVIFP.

The IVIFPs $\langle M_{k_i,c_j,d_s}, N_{k_i,c_j,d_s} \rangle$ are calculated on the basis of expert knowledge. The expert d_s $(s \in D)$ is considered unconscientious [7] if among his estimations $\langle M_{k_i,c_j,d_s}, N_{k_i,c_j,d_s} \rangle$ $(1 \le i \le m, 1 \le j \le n)$, there are estimations $\langle M_{k_i,w,l_j,w,d_s}, N_{k_i,w,l_j,w,d_s} \rangle$ $(w \in W, |W| \le mn)$ that are incorrect:

$$M_{k_i,w,l_j,w,d_s} \subseteq [0, 1], N_{k_i,w,l_j,w,d_s} \subseteq [0, 1], \sup M_{k_i,w,l_j,w,d_s} + \sup N_{k_i,w,l_j,w,d_s} \ge 1.$$

In [5, 7, 10], different ways for altering incorrect experts' estimations are discussed. We present one of them from [7]:

We replace M_{k_i,w,l_j,w,d_s} and N_{k_i,w,l_j,w,d_s} for $w \in W$, respectively, with $\overline{M_{k_i,w,l_j,w,d_s}}$ and $\overline{N_{k_i,w,l_j,w,d_s}}$ so that:

$$\inf \overline{M_{k_i,w,l_j,w,d_s}} = \inf M_{k_i,w,l_j,w,d_s} - \frac{\min(\inf M_{k_i,w,l_j,w,d_s}, \inf N_{k_i,w,l_j,w,d_s})}{2},$$

$$\sup \overline{M_{k_i,w,l_j,w,d_s}} = \sup M_{k_i,w,l_j,w,d_s} - \frac{\min(\sup M_{k_i,w,l_j,w,d_s}, \sup N_{k_i,w,l_j,w,d_s})}{2},$$

$$\inf \overline{N_{k_i,w,l_j,w,d_s}} = \inf N_{k_i,w,l_j,w,d_s} - \frac{\min(\inf M_{k_i,w,l_j,w,d_s}, \inf N_{k_i,w,l_j,w,d_s})}{2},$$

$$\sup \overline{N_{k_i,w,l_j,w,d_s}} = \sup N_{k_i,w,l_j,w,d_s} - \frac{\min(\sup M_{k_i,w,l_j,w,d_s}, \sup N_{k_i,w,l_j,w,d_s})}{2}.$$

In [7] a way was discussed for constructing of experts' scores. If the experts have a rating respectively $\{R_1, \ldots, R_s, \ldots, R_D\}$ and $R_s (1 \le s \le D)$ are IVIFPs, it is necessary before the aggregate evaluation of the candidates to perform the following operations in order to obtain a rating matrix that takes the ratings:

In the case of an optimistic forecast for the experts' scores:

$$ES^*[K, C, D, \{es^*_{k_i,c_j,d_s}\}]$$

$$= R_1 pr_{K,C,d_1} ES \oplus_{(max,min)} R_2 pr_{K,C,d_2} ES \oplus_{(max,min)} \cdots \oplus_{(max,min)} R_D pr_{K,C,d_D} ES.$$

In the case of a pessimistic forecast for the experts' scores:

$$ES^*[K, C, D, \{es^*_{k_i,c_j,d_s}\}]$$

$$= R_1 pr_{K,C,d_1} ES \oplus_{(min,max)} R_2 pr_{K,C,d_2} ES \oplus_{(min,max)} \cdots \oplus_{(min,max)} R_D pr_{K,C,d_D} ES.$$

Then,

$$ES := ES^*(es_{k_i,c_j,d_s} = es^*_{k_i,c_j,d_s}, \ \forall k_i \in K, \forall c_j \in C, \forall d_s \in D).$$

Let us apply the α_D-th aggregation operation (3) to find the aggregate value of the k_i-th candidate against the c_j-th criterion in a time-moment $h_g \notin D$ (for $1 \le i \le m$, $1 \le j \le n$, $1 \le g \le f$) as follows:

$$\alpha_{D,\#_q}(D, h_g) =
\begin{array}{c|ccc}
h_g & c_1 & \cdots & c_n \\
\hline
k_1 & \overset{D}{\underset{s=1}{\#_q}} \langle M_{k_1,l_1,h_g}, N_{k_1,l_1,h_g} \rangle & \cdots & \overset{D}{\underset{s=1}{\#_q}} \langle M_{k_1,l_n,h_g}, N_{k_1,l_n,h_g} \rangle \\
\vdots & \vdots & \ddots & \vdots \\
k_i & \overset{D}{\underset{s=1}{\#_q}} \langle M_{k_i,l_1,h_g}, N_{k_i,l_1,h_g} \rangle & \cdots & \overset{D}{\underset{s=1}{\#_q}} \langle M_{k_i,l_n,h_g}, N_{k_i,l_n,h_g} \rangle \\
\vdots & \vdots & \ddots & \vdots \\
k_m & \overset{D}{\underset{s=1}{\#_q}} \langle M_{k_m,l_1,h_g}, N_{k_m,l_1,h_g} \rangle & \cdots & \overset{D}{\underset{s=1}{\#_q}} \langle M_{k_m,l_n,h_g}, N_{k_m,l_n,h_g} \rangle
\end{array}$$

where $1 \le q \le 3$. If q is closer to the number 3, then the candidate's vacancy rating is more optimistic in a time-moment h_g.

Let us create the 3D-IVIFIM $A[K, C, H, \{a_{k_i,c_j,h_g}\}]$
$= \alpha_{D,\#_q}(ES, h_1) \oplus_{(max,min)} \alpha_{D,\#_q}(ES, h_2) \oplus_{(max,min)} \cdots \oplus_{(max,min)} \alpha_{D,\#_q}(ES, h_f)$

$$=
\begin{array}{c|ccccc}
h_g \in D & c_1 & \cdots & c_j & \cdots & c_n \\
\hline
k_1 & a_{k_1,c_1,h_g} & \cdots & a_{k_1,c_j,h_g} & \cdots & a_{k_1,c_n,h_g} \\
\vdots & \vdots & \ddots & \vdots & \ddots & \vdots \\
k_m & a_{k_m,c_1,h_g} & \cdots & a_{k_m,c_j,h_g} & \cdots & a_{k_m,c_n,h_g}
\end{array} ,$$

where $1 \le q \le 3$, $K = \{k_1, k_2, \ldots, k_m\}$, $C = \{c_1, c_2, \ldots, c_n\}$, $H = \{h_1, h_2, \ldots h_f\}$ and for $1 \le i \le m$, $1 \le j \le n$, $1 \le g \le f$: $\{a_{k_i,c_j,h_g}\} = \langle M_{k_i,c_j,h_g}, N_{k_i,c_j,h_g} \rangle$ is the IVIF-estimate of the k_i-th candidate for the c_j-th criterion in a time-moment h_g. If $\{a_{k_i,c_j,h_g}\}$ is an empty cell, then $\{a_{k_i,c_j,h_g}\} = \langle [0,0], [1,1] \rangle$, i.e., it is an interval-valued false pair.

We apply the aggregation operation by the dimension H to find the aggregated evelutions of the k_i-th candidate for the c_j-th criterion for the whole period in the organization (if one has not worked in the organization, one's aggregate score is the same as that applied for one's application for employment):

$$\alpha_{H,\#_q}(A, h_0) = \begin{array}{c|ccc} h_0 & c_1 & \cdots & c_n \\ \hline k_1 & \overset{D}{\underset{s=1}{\#_q}} \langle M_{k_1,l_1,h_g}, N_{k_1,l_1,h_g} \rangle & \cdots & \overset{D}{\underset{s=1}{\#_q}} \langle M_{k_1,l_n,h_g}, N_{k_1,l_n,h_g} \rangle \\ \vdots & \vdots & \ddots & \vdots \\ k_i & \overset{D}{\underset{s=1}{\#_q}} \langle M_{k_i,l_1,h_g}, N_{k_i,l_1,h_g} \rangle & \cdots & \overset{D}{\underset{s=1}{\#_q}} \langle M_{k_i,l_n,h_g}, N_{k_i,l_n,h_g} \rangle \\ \vdots & \vdots & \ddots & \vdots \\ k_m & \overset{D}{\underset{s=1}{\#_q}} \langle M_{k_m,l_1,h_g}, N_{k_m,l_1,h_g} \rangle & \cdots & \overset{D}{\underset{s=1}{\#_q}} \langle M_{k_m,l_n,h_g}, N_{k_m,l_n,h_g} \rangle \end{array}$$

where $h_0 \notin H$.

If q is closer to the number 3, then the evaluation of the candidates is more optimistic for the entire surveyed period.

Let us define the 2D-IVIFIM PK of the weight coefficients of the assessment criterion according to its priority to the corresponding position v_e $(1 \le e \le u)$

$$PK[C, V, \{pk_{c_j,v_e}\}] = \begin{array}{c|ccccc} & v_1 & \cdots & v_e & \cdots & v_u \\ \hline c_1 & pk_{c_1,v_1} & \cdots & pk_{c_1,v_e} & \cdots & pk_{c_1,v_u} \\ \vdots & \vdots & \ddots & \vdots & \ddots & \vdots \\ c_j & pk_{c_j,v_1} & \cdots & pk_{c_j,v_e} & \cdots & pk_{c_j,v_u} \\ \vdots & \vdots & \ddots & \vdots & \ddots & \vdots \\ c_n & pk_{c_n,v_1} & \cdots & pk_{c_n,v_e} & \cdots & pk_{c_n,v_u} \end{array},$$

where $C = \{c_1, c_2, \ldots, c_n\}$, $V = \{v_1, v_2, \ldots, v_u\}$ and for $1 \le j \le n, 1 \le e \le u$: pk_{c_j,v_e} are IVIFPs. Then we create 2D-IVIFIM $B[K, V, \{b_{k_i,v_e}\}]$:

$$B = \alpha_{H,\#_q}(A, h_0) \odot_{(\circ,*)} PK,$$

which contains the cumulative estimates of the k_i-th candidate (for $1 \le i \le m$) for the v_e-th vacancy (for $1 \le e \le u$) and $\langle \circ, * \rangle$ are the same as (3).

After this operation, we apply the aggregation operation $\alpha_{K,\#_q}(B, k_0)$ by the dimension K to find the most suitable candidate for the vacant position v_e.

$$\alpha_{K,\#_q}(B, k_0)$$

$$= \begin{array}{c|ccc} & v_1 & \cdots & v_u \\ \hline k_0 & \overset{m}{\underset{i=1}{\#_q}} \langle M_{k_i,v_1}, N_{k_i,v_1} \rangle & \cdots & \overset{m}{\underset{i=1}{\#_q}} \langle M_{k_i,v_u}, N_{k_i,v_u} \rangle \end{array},$$

where $k_0 \notin K$ and $1 \le q \le 3$.

After applying the interval-valued intuitionistic fuzzy Hungarian algorithm [25] with the cost IVIFIM B, we will get the optimal allocation of the candidates to the jobs. Prior to finding the optimal allocation of candidates to the vacancies, we can use one of the level-operators to IVIFIMs, setting a threshold for job applicants [10, 14].

4 An Example of the Application of the 3D-Interval-Valued Intuitionistic Fuzzy Appointment Problem

Let us consider the following problem (4) as an application of the algorithm, presented in Sect. 3:

We suppose that an organization wants to appoint (reappoint) the candidates $k_1, \ldots, k_4$ to positions $v_1, \ldots, v_4$. It is assumed that a 3-member committee d_1, d_2, d_3 scored the candidates and their assessments by the Munro Fraser's Five-Fold Grading System criteria $c_1, \ldots, c_5$ (1958, [15]) are IVIFPs. The aim of the problem is to optimally allocate vacancies among the candidates for them.   (4)

Let us introduce the set of criteria $C = \{c_1, c_2, c_3, c_4, c_5\}$, which present the five-fold person specification grading:

- c_1—Impact on others by way of appearance and expression
- c_2—Job-related training and experience acquired over years
- c_3—In-born qualities that blend with the ability to grasp things quickly
- c_4—Motivation to carry out assigned tasks
- c_5—Capability of coping with stress.

Let us create the following 3D-IVIFIM, in accordance with the problem (4):

$$ES[K, C, D, \{es_{k_i, c_j, d_s}\}]$$

$$= \begin{cases} \begin{array}{c|ccc} d_1 & c_1 & c_2 & c_3 \\ \hline k_1 & \langle[0.4, 0.5], [0.3; 0.4]\rangle & \langle[0.7, 0.8], [0.1, 0.2]\rangle & \langle[0.3, 0.4], [0.2, 0.3]\rangle \\ k_2 & \langle[0.7, 0.8], [0.1, 0.2]\rangle & \langle[0.5, 0.6], [0.1, 0.2]\rangle & \langle[0.5, 0.6], [0.2, 0.3]\rangle \\ k_3 & \langle[0.5, 0.6], [0.2, 0.3]\rangle & \langle[0.6, 0.7], [0.1, 0.2]\rangle & \langle[0.6, 0.7], [0.2, 0.3]\rangle \\ k_4 & \langle[0.6, 0.7], [0.0, 0.1]\rangle & \langle[0.7, 0.8], [0.1, 0.2]\rangle & \langle[0.8, 0.9], [0.0, 0.1]\rangle \end{array} \end{cases}$$

c_4	c_5
$\langle[0.5, 0.6], [0.1, 0.2]\rangle$	$\langle[0.6, 0.7], [0.0, 0.1]\rangle$
$\langle[0.3, 0.4], [0.4, 0.5]\rangle$	$\langle[0.7, 0.8], [0.0, 0.1]\rangle$,
$\langle[0.2, 0.3], [0.4, 0.5]\rangle$	$\langle[0.6, 0.7], [0.1, 0.2]\rangle$
$\langle[0.4, 0.5], [0.3, 0.4]\rangle$	$\langle[0.5, 0.6][0.2, 0.3]\rangle$

d_2	c_1	c_2	c_3
k_1	$\langle[0.3, 0.4], [0.2, 0.3]\rangle$	$\langle[0.9, 1], [0.1, 0.0]\rangle$	$\langle[0.4, 0.5], [0.3, 0.4]\rangle$
k_2	$\langle[0.6, 0.7], [0.1, 0.2]\rangle$	$\langle[0.4, 0.5], [0.2, 0.3]\rangle$	$\langle[0.4, 0.5], [0.0, 0.1]\rangle$
k_3	$\langle[0.4, 0.5], [0.3, 0.4]\rangle$	$\langle[0.7, 0.8], [0.0, 0.1]\rangle$	$\langle[0.5, 0.6], [0.2, 0.3]\rangle$
k_4	$\langle[0.5, 0.6], [0.0, 0.1]\rangle$	$\langle[0.6, 0.7], [0.1, 0.2]\rangle$	$\langle[0.7, 0.8], [0.0, 0.1]\rangle$

c_4	c_5
$\langle[0.4, 0.5], [0.0, 0.1]\rangle$	$\langle[0.5, 0.6], [0.1, 0.2]\rangle$
$\langle[0.4, 0.5], [0.3, 0.4]\rangle$	$\langle[0.6, 0.7], [0.0, 0.1]\rangle$,
$\langle[0.2, 0.3], [0.5, 0.6]\rangle$	$\langle[0.7, 0.8], [0.1, 0.2]\rangle$
$\langle[0.5, 0.6], [0.1, 0.2]\rangle$	$\langle[0.3, 0.4], [0.2, 0.3]\rangle$

d_3	c_1	c_2	c_3
k_1	$\langle[0.5, 0.6], [0.3, 0.4]\rangle$	$\langle[0.6, 0.7], [0.1, 0.2]\rangle$	$\langle[0.2, 0.3], [0.3, 0.4]\rangle$
k_2	$\langle[0.7, 0.8], [0.0, 0.1]\rangle$	$\langle[0.6, 0.7], [0.1, 0.2]\rangle$	$\langle[0.8, 0.7], [0.0, 0.1]\rangle$
k_3	$\langle[0.6, 0.7], [0.1, 0.2]\rangle$	$\langle[0.7, 0.8], [0.1, 0.2]\rangle$	$\langle[0.7, 0.8], [0.0, 0.1]\rangle$
k_4	$\langle[0.8, 0.9], [0.0, 0.1]\rangle$	$\langle[0.4, 0.5], [0.3, 0.4]\rangle$	$\langle[0.6, 0.7], [0.0, 0.1]\rangle$

c_4	c_5
$\langle[0.2, 0.3], [0.5, 0.6]\rangle$	$\langle[0.6, 0.7], [0.1, 0.2]\rangle$
$\langle[0.1, 0.2], [0.4, 0.5]\rangle$	$\langle[0.4, 0.5], [0.0, 0.1]\rangle$,
$\langle[0.2, 0.3], [0.3, 0.4]\rangle$	$\langle[0.7, 0.8], [0.1, 0.2]\rangle$
$\langle[0.6, 0.7], [0.1, 0.2]\rangle$	$\langle[0.4, 0.5], [0.2, 0.3]\rangle$

where $K = \{k_1, k_2, k_3, k_4\}$, $C = \{c_1, c_2, c_3, c_4, c_5\}$, $D = \{d_1, d_2, d_3\}$ and the element $\{es_{k_i,c_j,d_s}\} = \langle M_{k_i,c_j,d_s}, N_{k_i,c_j,d_s}\rangle$ (for $1 \leq i \leq 4, 1 \leq j \leq 5, 1 \leq s \leq 3$) is the estimate of the d_s-th expert for the k_i-th employee by the c_j-th criterion and that is an IVIFP.

If the experts have the following rating coefficients, respectively,

$$\{R_1, R_2, R_3\} = \{\langle[0.7, 0.8], [0.0, 0.1]\rangle, \langle[0.6, 0.7], [0.0, 0.1]\rangle, \langle[0.8, 0.9], [0.0, 0.1]\rangle\},$$

it is necessary to perform the following operations before the aggregate evaluation of the candidates, in order to obtain a rating matrix that takes the ratings. We take the optimistic forecast, then we use the software "IndMatCalc v 0.9" [23] to implement the matrix operations:

$$ES^*[K, C, D, \{es^*_{k_i,c_j,d_s}\}]$$

$$= R_1 pr_{K,C,d_1} ES \oplus_{(max,min)} R_2 pr_{K,C,d_2} ES \oplus_{(max,min)} R_3 pr_{K,C,d_3} ES.$$

d_1	c_1	c_2	c_3
k_1	$\langle[0.28, 0.4], [0.3; 0.46]\rangle$	$\langle[0.49, 0.64], [0.1, 0.28]\rangle$	$\langle[0.21, 0.32], [0.2, 0.37]\rangle$
k_2	$\langle[0.49, 0.64], [0.1, 0.28]\rangle$	$\langle[0.35, 0.48], [0.1, 0.28]\rangle$	$\langle[0.35, 0.48], [0.2, 0.37]\rangle$
k_3	$\langle[0.35, 0.48], [0.2, 0.37]\rangle$	$\langle[0.42, 0.56], [0.1, 0.28]\rangle$	$\langle[0.42, 0.56], [0.2, 0.37]\rangle$
k_4	$\langle[0.42, 0.56], [0.0, 0.19]\rangle$	$\langle[0.49, 0.64], [0.1, 0.28]\rangle$	$\langle[0.56, 0.72], [0.0, 0.19]\rangle$

	c_4	c_5
	$\langle[0.35,0.48],[0.1,0.28]\rangle$	$\langle[0.42,0.56],[0.0,0.19]\rangle$
	$\langle[0.21,0.32],[0.4,0.55]\rangle$	$\langle[0.49,0.64],[0.0,0.19]\rangle,$
	$\langle[0.14,0.24],[0.4,0.55]\rangle$	$\langle[0.42,0.56],[0.1,0.28]\rangle$
	$\langle[0.28,0.4],[0.3,0.46]\rangle$	$\langle[0.35,0.48][0.2,0.37]\rangle$

d_2	c_1	c_2	c_3
k_1	$\langle[0.18,0.28],[0.2,0.37]\rangle$	$\langle[0.54,0.7],[0.1,0.1]\rangle$	$\langle[0.24,0.35],[0.3,0.46]\rangle$
k_2	$\langle[0.36,0.49],[0.1,0.28]\rangle$	$\langle[0.24,0.35],[0.2,0.37]\rangle$	$\langle[0.24,0.35],[0.0,0.19]\rangle$
k_3	$\langle[0.24,0.35],[0.3,0.46]\rangle$	$\langle[0.42,0.56],[0.0,0.19]\rangle$	$\langle[0.3,0.42],[0.2,0.37]\rangle$
k_4	$\langle[0.3,0.42],[0.0,0.19]\rangle$	$\langle[0.36,0.49],[0.1,0.28]\rangle$	$\langle[0.42,0.56],[0.0,0.19]\rangle$

	c_4	c_5
	$\langle[0.3,0.35],[0.1,0.19]\rangle$	$\langle[0.18,0.42],[0.2,0.28]\rangle$
	$\langle[0.24,0.35],[0.0,0.46]\rangle$	$\langle[0.3,0.49],[0.1,0.19]\rangle$,
	$\langle[0.24,0.21],[0.3,0.64]\rangle$	$\langle[0.36,0.56],[0.0,0.28]\rangle$
	$\langle[0.12,0.42],[0.1,0.28]\rangle$	$\langle[0.18,0.28],[0.2,0.37]\rangle$

d_3	c_1	c_2	c_3
k_1	$\langle[0.4,0.54],[0.3,0.46]\rangle$	$\langle[0.48,0.63],[0.1,0.28]\rangle$	$\langle[0.16,0.27],[0.3,0.46]\rangle$
k_2	$\langle[0.56,0.72],[0.0,0.19]\rangle$	$\langle[0.48,0.63],[0.1,0.28]\rangle$	$\langle[0.48,0.63],[0.0,0.19]\rangle$
k_3	$\langle[0.48,0.63],[0.10,0.28]\rangle$	$\langle[0.56,0.72],[0.1,0.28]\rangle$	$\langle[0.56,0.72],[0.0,0.18]\rangle$
k_4	$\langle[0.64,0.81],[0.0,0.19]\rangle$	$\langle[0.32,0.45],[0.3,0.46]\rangle$	$\langle[0.48,0.63],[0.0,0.19]\rangle$

	c_4	c_5
	$\langle[0.16,0.27],[0.5,0.63]\rangle$	$\langle[0.48,0.63],[0.1,0.28]\rangle$
	$\langle[0.08,0.18],[0.4,0.55]\rangle$	$\langle[0.32,0.45],[0.0,0.19]\rangle$
	$\langle[0.16,0.27],[0.3,0.46]\rangle$	$\langle[0.56,0.72],[0.1,0.28]\rangle$
	$\langle[0.48,0.63],[0.1,0.28]\rangle$	$\langle[0.32,0.45],[0.2,0.37]\rangle$

Then,

$$ES := ES^*(es_{k_i,c_j,d_s} = es^*_{k_i,c_j,d_s}, \ \forall k_i \in K, \forall c_j \in C, \forall d_s \in D).$$

Let us apply the optimistic aggregation operation $\alpha_{D,\#_3}$ to find the aggregate value of the k_i-th candidate against the c_j-th criterion in a current time-moment $h_3 \notin D$ (for $1 \le i \le 4, 1 \le j \le 5$) as follows:

h_3	c_1	c_2
k_1	$\langle[0.4,0.54],[0.3,0.37]\rangle$	$\langle[0.54,0.7],[0.1,0.1]\rangle$
k_2	$\langle[0.56,0.72],[0.0,0.19]\rangle$	$\langle[0.48,0.63],[0.1,0.28]\rangle$
k_3	$\langle[0.48,0.63],[0.1,0.28]\rangle$	$\langle[0.42,0.72],[0.0,0.19]\rangle$
k_4	$\langle[0.64,0.81],[0.0,0.19]\rangle$	$\langle[0.49,0.64],[0.1,0.28]\rangle$

$\alpha_{D,\#_3}(ES, h_3) = $

	c_3	c_4	c_5
	$\langle[0.24, 0.35], [0.2, 0.37]\rangle$	$\langle[0.35, 0.48], [0.1, 0.19]\rangle$	$\langle[0.35, 0.48], [0.2, 0.37]\rangle$
	$\langle[0.48, 0.63], [0.0, 0.19]\rangle$	$\langle[0.24, 0.35], [0.0, 0.46]\rangle$	$\langle[0.49, 0.64], [0.0, 0.19]\rangle$.
	$\langle[0.56, 0.72], [0.0, 0.18]\rangle$	$\langle[0.24, 0.27], [0.3, 0.46]\rangle$	$\langle[0.56, 0.72], [0.0, 0.28]\rangle$
	$\langle[0.56, 0.72], [0.0, 0.19]\rangle$	$\langle[0.48, 0.63], [0.1, 0.28]\rangle$	$\langle[0.35, 0.48][0.0, 0.37]\rangle$

Let us suppose that staff rating on the criteria has existed for 3 years. Employee evaluation data has been stored for the two previous periods h_1 and h_2. Then, we create the 3D-IVIFIM

$$A[K, C, H, \{a_{k_i, c_j, h_g}\}]$$

$$= \alpha_{D, \#_3}(ES, h_1) \oplus_{(max, min)} \alpha_{D, \#_3}(ES, h_2) \oplus_{(max, min)} \cdots \oplus_{(max, min)} \alpha_{D, \#_3}(ES, h_f)$$

h_1	c_1	c_2
k_1	$\langle[0.18, 0.36], [0.3, 0.46]\rangle$	$\langle[0.54, 0.63], [0.1, 0.28]\rangle$
k_2	$\langle[0.36, 0.48], [0.2, 0.28]\rangle$	$\langle[0.48, 0.54], [0.1, 0.24]\rangle$
k_3	$\langle[0.40, 0.60], [0.2, 0.42]\rangle$	$\langle[0.42, 0.56], [0.1, 0.28]\rangle$
k_4	$\langle[0.0, 0.0], [1.0, 1.0]\rangle$	$\langle[0.0, 0.0], [1.0, 1.0]\rangle$

c_3	c_4	c_5
$\langle[0.16, 0.27], [0.3, 0.46]\rangle$	$\langle[0.16, 0.27], [0.1, 0.63]\rangle$	$\langle[0.18, 0.42], [0.2, 0.28]\rangle$
$\langle[0.24, 0.35], [0.2, 0.37]\rangle$	$\langle[0.72, 0.81], [0.1, 0.19]\rangle$	$\langle[0.63, 0.72], [0.1, 0.19]\rangle$,
$\langle[0.3, 0.42], [0.2, 0.37]\rangle$	$\langle[0.14, 0.21], [0.4, 0.64]\rangle$	$\langle[0.36, 0.56], [0.1, 0.28]\rangle$
$\langle[0.0, 0.0], [1.0, 1.0]\rangle$	$\langle[0.0, 0.0], [1.0, 1.0]\rangle$	$\langle[0.0, 0.0], [1.0, 1.0]\rangle$

h_2	c_1	c_2
k_1	$\langle[0.46, 0.56], [0.1, 0.2]\rangle$	$\langle[0.42, 0.63], [0.1, 0.2]\rangle$
k_2	$\langle[0.54, 0.64], [0.1, 0.28]\rangle$	$\langle[0.45, 0.56], [0.1, 0.28]\rangle$
k_3	$\langle[0.54, 0.63], [0.1, 0.2]\rangle$	$\langle[0.72, 0.81], [0.1, 0.19]\rangle$
k_4	$\langle[0.0, 0.0], [1.0, 1.0]\rangle$	$\langle[0.0, 0.0], [1.0, 1.0]\rangle$

c_3	c_4	c_5
$\langle[0.42, 0.63], [0.3, 0.46]\rangle$	$\langle[0.49, 0.56], [0.1, 0.2]\rangle$	$\langle[0.56, 0.63], [0.2, 0.28]\rangle$
$\langle[0.54, 0.63], [0.1, 0.2]\rangle$	$\langle[0.27, 0.36], [0.2, 0.3]\rangle$	$\langle[0.72, 0.81], [0.1, 0.19]\rangle$,
$\langle[0.36, 0.54], [0.2, 0.37]\rangle$	$\langle[0.36, 0.48], [0.1, 0.2]\rangle$	$\langle[0.42, 0.54], [0.1, 0.28]\rangle$
$\langle[0.0, 0.0], [1.0, 1.0]\rangle$	$\langle[0.0, 0.0], [1.0, 1.0]\rangle$	$\langle[0.0, 0.0], [1.0, 1.0]\rangle$

h_3	c_1	c_2
k_1	$\langle[0.4, 0.54], [0.3, 0.37]\rangle$	$\langle[0.54, 0.7], [0.1, 0.1]\rangle$
k_2	$\langle[0.56, 0.72], [0.0, 0.19]\rangle$	$\langle[0.48, 0.63], [0.1, 0.28]\rangle$
k_3	$\langle[0.48, 0.63], [0.1, 0.28]\rangle$	$\langle[0.42, 0.72], [0.0, 0.19]\rangle$
k_4	$\langle[0.64, 0.81], [0.0, 0.19]\rangle$	$\langle[0.49, 0.64], [0.1, 0.28]\rangle$

$$
\left.
\begin{array}{ccc}
c_3 & c_4 & c_5 \\
\hline
\langle[0.24, 0.35], [0.2, 0.37]\rangle & \langle[0.35, 0.48], [0.1, 0.19]\rangle & \langle[0.35, 0.48], [0.2, 0.37]\rangle \\
\langle[0.48, 0.63], [0.0, 0.19]\rangle & \langle[0.24, 0.35], [0.0, 0.46]\rangle & \langle[0.49, 0.64], [0.0, 0.19]\rangle \\
\langle[0.56, 0.72], [0.0, 0.18]\rangle & \langle[0.24, 0.27], [0.3, 0.46]\rangle & \langle[0.48, 0.72], [0.0, 0.28]\rangle \\
\langle[0.56, 0.72], [0.0, 0.19]\rangle & \langle[0.48, 0.63], [0.1, 0.28]\rangle & \langle[0.35, 0.48][0.0, 0.37]\rangle
\end{array}
\right\},
$$

where $K = \{k_1, k_2, k_3, k_4\}$, $C = \{c_1, c_2, c_3, c_4, c_5\}$, $H = \{h_1, h_2, h_3\}$, $\{a_{k_i, c_j, h_s}\}$ (for $1 \le i \le 4, 1 \le j \le 5, 1 \le s \le 3$) is the estimate of the k_i-th candidate by the c_j-th criterion in a time-moment h_g and that is an IVIFP.

After application of the optimistic aggregation operation $\alpha_{H, \#_3}$ by the dimension H, we find the aggregated evalutions of the k_i-th candidate for the c_j-th criterion for the entire three-year period in which one worked (we assume that the company has been established three years ago) or is still working in the organization. In our example, if the candidate k_4 has not worked in the organization, then one's aggregate score is the same as that applied for the application for employment:

$$
\alpha_{H, \#_3}(A, h_0) =
\begin{array}{c|cc}
h_0 & c_1 & c_2 \\
\hline
k_1 & \langle[0.46, 0.56], [0.1, 0.2]\rangle & \langle[0.54, 0.7], [0.1, 0.1]\rangle \\
k_2 & \langle[0.56, 0.72], [0.0, 0.19]\rangle & \langle[0.48, 0.63], [0.1, 0.24]\rangle \\
k_3 & \langle[0.54, 0.63], [0.1, 0.2]\rangle & \langle[0.72, 0.81], [0.0, 0.19]\rangle \\
k_4 & \langle[0.64, 0.81], [0.0, 0.19]\rangle & \langle[0.49, 0.64], [0.1, 0.28]\rangle
\end{array}
$$

$$
\begin{array}{ccc}
c_3 & c_4 & c_5 \\
\hline
\langle[0.42, 0.63], [0.2, 0.37]\rangle & \langle[0.49, 0.56], [0.1, 0.19]\rangle & \langle[0.56, 0.63], [0.2, 0.28]\rangle \\
\langle[0.54, 0.63], [0.0, 0.19]\rangle & \langle[0.72, 0.81], [0.0, 0.19]\rangle & \langle[0.72, 0.81], [0.0, 0.19]\rangle, \\
\langle[0.56, 0.72], [0.0, 0.18]\rangle & \langle[0.36, 0.48], [0.1, 0.2]\rangle & \langle[0.48, 0.72], [0.0, 0.28]\rangle \\
\langle[0.56, 0.72], [0.0, 0.19]\rangle & \langle[0.48, 0.63], [0.1, 0.28]\rangle & \langle[0.35, 0.48], [0.0, 0.37]\rangle
\end{array}
$$

where $h_0 \notin H$. We create 2D-IVIFIM PK of the weight coefficients of the assessment criteria according to their priority to the corresponding position v_e ($1 \le e \le 4$)

$$
PK[C, V, \{pk_{c_j, v_e}\}] =
\begin{array}{c|cc}
 & v_1 & v_2 \\
\hline
c_1 & \langle[0.8, 0.9], [0.0, 0.1]\rangle & \langle[0.7, 0.8], [0.1, 0.2]\rangle \\
c_2 & \langle[0.7, 0.8], [0.0, 0.1]\rangle & \langle[0.5, 0.6], [0.1, 0.2]\rangle \\
c_3 & \langle[0.5, 0.6], [0.1, 0.2]\rangle & \langle[0.8, 0.9], [0.0, 0.1]\rangle \\
c_4 & \langle[0.8, 0.9], [0.0, 0.1]\rangle & \langle[0.8, 0.9], [0.0, 0.1]\rangle \\
c_5 & \langle[0.7, 0.8], [0.1, 0.2]\rangle & \langle[0.8, 0.9], [0.0, 0.1]\rangle
\end{array}
$$

$$
\begin{array}{cc}
v_3 & v_4 \\
\hline
\langle[0.5, 0.6], [0.1, 0.2]\rangle & \langle[0.6, 0.7], [0.1, 0.2]\rangle \\
\langle[0.6, 0.7], [0.0, 0.2]\rangle & \langle[0.7, 0.8], [0.1, 0.2]\rangle \\
\langle[0.4, 0.5], [0.2, 0.3]\rangle & \langle[0.6, 0.7], [0.1, 0.2]\rangle \\
\langle[0.7, 0.8], [0.0, 0.1]\rangle & \langle[0.6, 0.7], [0.1, 0.2]\rangle \\
\langle[0.8, 0.9], [0.0, 0.1]\rangle & \langle[0.5, 0.6], [0.3, 0.4]\rangle
\end{array}
$$

where $C = \{c_1, c_2, \ldots, c_5\}$, $V = \{v_1, v_2, v_3, v_4\}$ and for $1 \le j \le 5, 1 \le e \le 4$: pk_{c_j, v_e} are IVIFPs.

Then we create 2D-IVIFIM $B[K, V, \{b_{k_i, v_e}\}]$:

$$B = \alpha_{H, \#_3}(A, h_0) \odot_{(max, min)} PK$$

	v_1	v_2
k_1	$\langle [0.5, 0.63], [0.1, 0.2] \rangle$	$\langle [0.54, 0.57], [0.1, 0.2] \rangle$
$= k_2$	$\langle [0.54, 0.63], [0.0, 0.1] \rangle$	$\langle [0.5, 0.63], [0.0, 0.1] \rangle$
k_3	$\langle [0.56, 0.7], [0.0, 0.1] \rangle$	$\langle [0.7, 0.8], [0.0, 0.1] \rangle$
k_4	$\langle [0.56, 0.72], [0.0, 0.1] \rangle$	$\langle [0.5, 0.64], [0.0, 0.1] \rangle$

v_3	v_4
$\langle [0.42, 0.63], [0.0, 0.1] \rangle$	$\langle [0.56, 0.63], [0.1, 0.1] \rangle$
$\langle [0.54, 0.63], [0.0, 0.1] \rangle$	$\langle [0.6, 0.7], [0.0, 0.19] \rangle$,
$\langle [0.54, 0.63], [0.0, 0.1] \rangle$	$\langle [0.5, 0.7], [0.0, 0.18] \rangle$
$\langle [0.56, 0.7], [0.0, 0.1] \rangle$	$\langle [0.5, 0.6], [0.0, 0.19] \rangle$

which contains the cumulative optimistic estimates of the k_i-th candidate (for $1 \le i \le 4$) for the v_e-th vacancy (for $1 \le e \le 4$). After this operation, we apply the aggregation operation $\alpha_{H, \#_3}(B, k_0)$ by the dimension K, to find the most suitable candidate for the vacant position v_e with the maximum criteria rating:

$$\alpha_{K, \#_3}(B, k_0)$$

	v_1	v_2
$= k_0$	$\langle [0.56, 0.72], [0.0, 0.1] \rangle$	$\langle [0.7, 0.8], [0.0, 0.1] \rangle$

v_3	v_4
$\langle [0.56, 0.7], [0.0, 0.1] \rangle$	$\langle [0.6, 0.7], [0.0, 0.19] \rangle$

where $k_0 \notin K$. We get the following optimal allocation of candidates for positions: candidate k_1 for position v_3, candidate k_2—for position v_4, candidate k_3—for position v_2 and candidate k_4—for position v_1.

5 Conclusion

In the paper was defined the three-dimensional interval-valued intuitionistic fuzzy appointment problem an algorithm was proposed for its optimal solution, in which the evaluations of candidates against the criteria set by several experts in a certain moment time are IVIFPs. The proposed algorithm for the solution takes into account the ratings of the experts and the weight coefficients of the assessment criteria according to their priorities for the respective position.

The outlined approach to supporting decision making in an uncertain environment, has the following advantages: it can be applied to both the appointment problem with crisp parameters and with intuitionistic fuzzy ones; the algorithm can be extended in order to obtain the optimal solution for other types of multidimensional appointment problems.

The proposed example demonstrates the correctness and efficiency of the algorithm. In future, the proposed method will be implemented for various types of multidimensional optimal problems with fuzzy or intuitionistic fuzzy data.

Acknowledgements This work was supported by the Bulgarian Ministry of Education and Science under the National Research Programme "Young scientists and postdoctoral students", approved by DCM # 577/ 17.08.2018.

References

1. Atanassov, K.: Generalized index matrices. Comptes rendus de l'Academie Bulgare des Sciences **40**(11), 15–18 (1987)
2. Atanassov, K.: Review and new results on intuitionistic fuzzy sets. Preprint IM-MFAIS-1-88, Sofia (1988)
3. Atanassov, K.: Generalized Nets. World Scientific, Singapore (1991)
4. Atanassov, K.: Operators over interval valued intuitionistic fuzzy sets. Fuzzy Sets Syst. **64**(2), 159–174 (1994)
5. Atanassov, K.: Intuitionistic Fuzzy Sets. Springer, Heidelberg (1999)
6. Atanassov, K.: On Generalized Nets Theory. "Prof. M. Drinov". Academic Publishing House, Sofia (2007)
7. Atanassov, K.: On Intuitionistic Fuzzy Sets Theory. STUDFUZZ, vol. 283. Springer, Heidelberg (2012). https://doi.org/10.1007/978-3-642-29127-2
8. Atanassov, K.: Index Matrices: Towards an Augmented Matrix Calculus. Studies in Computational Intelligence, vol. 573. Springer, Cham (2014). https://doi.org/10.1007/978-3-319-10945-9
9. Atanassov, K.: Extended interval valued intuitionistic fuzzy index matrices. (submitted to Springer) (2019). (in press)
10. Atanassov, K.: Interval valued intuitionistic fuzzy sets. (submitted to Springer) (2019). (in press)
11. Atanassov, K., Gargov, G.: Interval valued intuitionistic fuzzy sets. Fuzzy Sets Syst. **31**(3), 343–349 (1989)
12. Atanassov, K., Vassilev, P., Kacprzyk, J., Szmidt, E.: On interval valued intuitionistic fuzzy pairs. J. Univ. Math. **1**(3), 261–268 (2018)
13. Atanassov, K.T.: Intuitionistic fuzzy sets, VII ITKR Session, Sofia, 20–23 June 1983 (Deposed in Centr. Sci.-Techn. Library of the Bulg. Acad. of Sci., 1697/84) (in Bulgarian). Reprinted: Int. J. Bioautomation **20**(S1), S1–S6 (2016)
14. Atanassova, V.: New modified level operator N_γ over intuitionistic fuzzy sets. In: Proceedings of 12th International Conference on Flexible Query Answering Systems (FQAS 2017), London, UK, 21–22 June 2017; (Christiansen, H., Jaudoin, H., Chountas, P., Andreasen, T., Larsen, H.L. (eds.) LNAI, vol. 10333. Springer, pp. 209–214 (2017). https://doi.org/10.1007/978-3-319-59692-1_18
15. Bratton, J., Gold, J.: Human Resource Management: Theory and Practice, 5th edn. Macmillan press, London (2012)

16. Bureva, V., Sotirova, E., Moskova, M., Szmidt, E.: Solving the problem of appointments using index matrices. In: Proceedings of the 13th International Workshop on Generalized Nets, vol. 29, pp. 43–48 (2012)
17. De, S.K., Biswas, R., Roy, A.R.: Some operations on intuitionistic fuzzy sets. Fuzzy Sets Syst. **114**(4), 477–484 (2000)
18. Ejegwa, P.: Intuitionistic fuzzy sets approach in appointment of positions in an organization via max-min-max rule. Glob. J. Sci. Front. Res. F Math. Decis. Sci. **15**(6) (2015)
19. Ejegwa, P., Modom, E.: Diagnosis of viral hepatitis using new distance measure of intuitionistic fuzzy sets. Int. J. Fuzzy Math. Arch. **8**(1), 1–7 (2015)
20. Ejegwa P., Kwarkar, L., Ihuoma, K.N.: Application of intuitionistic fuzzy multisets in appointment process. Int. J. Comput. Appl. (0975 8887) **135**(1) (2016)
21. Gorzalczany, M.: Interval-valued fuzzy inference method - some basic properties. Fuzzy Sets Syst. **31**(2), 243–251 (1989)
22. Petkova-Georgieva, S.: Economy of healthcare. University Publishing Prof, Assen Zlatarov, Burgas, Bulgaria (2017). (in Bulgarian)
23. Software for index matrices. http://justmathbg.info/indmatcalc.html. Accessed 1 Feb 2019
24. Traneva, V., Tranev, S.: Index Matrices as a Tool for Managerial Decision Making. Publishing House of the Union of Scientists, Bulgaria (2017). (in Bulgarian)
25. Traneva, V., Tranev, S.: An interval valued intuitionistic fuzzy approach to assignment problem. In: Proceedings of Infus 2019 (2019). (submitted to Springer, Cham)
26. Traneva, V., Sotirova, E., Bureva, V., Atanassov, K.: Aggregation operations over 3-dimensional extended index matrices. Adv. Stud. Contemp. Math. **25**(3), 407–416 (2015)
27. Traneva, V., Tranev, S., Stoenchev, M., Atanassov, K.: Scaled aggregation operations over two- and three-dimensional index matrices. Soft Comput. **22**(15), 5115–5120 (2018). https://doi.org/10.1007/s00500-018-3315-6
28. Traneva, V., Atanassova, V., Tranev, S.: Index matrices as a decision-making tool for job appointment. In: Nikolov G., Kolkovska N., Georgiev K. (eds) Numerical Methods and Applications. NMA 2018. Lecture Notes in Computer Science, vol. 11189, 158–166. Springer, Cham (2019). https://doi.org/10.1007/978-3-030-10692-8_18
29. Zadeh, L.: Fuzzy sets. Inf. Control. **8**(3), 338–353 (1965)

Logical Connectives Used in the First Bulgarian School Books in Mathematics

Velislava Stoykova

Abstract The paper presents a statistical approach to extract and analyze logical connectives used in the first Bulgarian school books in mathematics published during the first half of XIX c. It applies Information Retrieval (IR) approach and the Sketch Engine software to search the electronic readable format of the books (without normalizing their graphical representation). The approach is based on the use of complex search techniques to evaluate related queries (keywords) and further, the query search is optimized by limiting the scope of the search using options according to several search criteria. The results are analyzed both with respect to the syntactic distribution (generally as adverbs and conjunctions) and to the semantics they express (generally as transitions).

1 Introduction

The contemporary formal syntactic theories regard the text taking into account its co-reference structure and related internal mechanisms to carry, maintain and transmit the information about its logical coherence. The co-referencing text relations are usually expressed by using related connectives, which are words (mainly conjunctions, prepositions and adverbs) or phrases that links clauses or sentences, so to maintain their semantic relations [4]. The connectives [9] can express relations like order, sequence, comparison, importance, etc. and are used to connect ideas, paragraphs and sentences.

At the same time, in the computer-based approaches to language, a formal syntactic annotations [13] and encodings are required, so to process (to parse) the electronic language resources in order to extract related connectives. In our research, we use a statistical approach instead, so to extract the logical connectives which are specific for the first mathematical texts written in Bulgarian language.

V. Stoykova (✉)
Institute for Bulgarian Language, Bulgarian Academy of Sciences,
52, Shipchensky proh. str., bl. 17, 1113 Sofia, Bulgaria
e-mail: vstoykova@yahoo.com

© Springer Nature Switzerland AG 2020
S. Fidanova (ed.), *Recent Advances in Computational Optimization*,
Studies in Computational Intelligence 838,
https://doi.org/10.1007/978-3-030-22723-4_13

2 Logical and Linguistic Connectives

Following its definition, the connectives are natural language words (or phrases) which function as connections between phrases and sentences and are regarded as expressing logical relations in a grammatically correct way. Thus, connectives can be analyzed both as linguistic and as logical markers [3, 7].

Logically, the connectives are the symbols (or words) used to connect two or more sentences in a grammatically valid way and the resultative meaning depends on that of the original sentence and on the meaning of the connective. That symbols (words) can present various types of logical relations.

Grammatically, the connectives typically are prepositions, adverbs and conjunctions. Also, commas, in certain syntactic positions (when connect adverb or adverbial modifier), can be regarded as connectives. Thus, connectives generally express a syntactic functions and can be extracted from the text following the syntactic rules which connect sentences (through conjunctions and prepositions).

The conjunctive adverbs function as a connection between two clauses that convert the clause they introduce into adverbial modifier. Often, the conjunctive adverbials modify the verb, the adjective, or another adverb in the main clause, and in that way, they modify the previously expressed logical predication. The conjunctive adverb functions as an adverbial connective (also known as a logical transition) which is used within a second clause, so to show its logical relationship to the first. That relations can represent sequence, contrast, cause and effect, purpose or reason, etc.

Some authors tend to analyze connectives at the text level introducing the term 'transitions' and defining them as providing the text cohesion by making text more explicit or by signaling how ideas relate to one another [16]. The transitions are divided into coordinating, subordinating, temporal, etc. Nevertheless, for both analyses of conjunctive adverbs (as *logical connectives* or as *transitions*) a formal syntactic analysis is required.

However, our approach is based on the use of the Distributional Semantic Model frameworks which regards the use of statistical similarity (or distance) as relevant for semantic similarity (or distance) [2]. Thus, we evaluate statistically-significant words as semantically-relevant and we extract most common words which function as typical for that mathematical texts connectives (mostly conjunctive adverbs) showing additional information about their variability and functionality.

The techniques used adopt metrics for extraction of different types of word semantic relations by estimation of word similarity measure [1]. Further, we are going to present such techniques and to discuss the received results of using statistical functions of the Sketch Engine software [12] for keywords extraction (keyness), concordances and collocations generation and semantic clustering in searching electronic educational resource which contains the first three mathematical school books published in Bulgarian language during the first half of XIX c.

3 Sketch Engine (SE)

The SE software allows approaches to extract semantic properties of words and most of them are with multilingual application. Generating keywords (*keyness*) is a widely used technique to extract terms of particular studied domain. Also, semantic relations can be extracted by generation of related word contexts through word concordances which define context in quantitative terms and a further work is needed to be done to extract semantic relations by searching for co-occurrences and collocations of related keyword.

Co-occurrences and collocations are words which are most probably to be found with a related keyword. They assign the semantic relations between the keyword and its particular collocated word which might be of a similarity or of a distance. The statistical approaches used by SE to search for co-occurrence and collocated words are based on defining the probability of their co-occurrences and collocations. We use techniques of *Tscore*, *MIscore* and *MI³score* for process and search corpora. For all, the following terms are used: N—corpus size, f_A—number of occurrences of keyword in the whole corpus (the size of concordance), f_B—number of occurrences of collocated keyword in the whole corpus, f_{AB}—number of occurrences of collocate in the concordance (number of co-occurrences). The related formulas for defining *Tscore*, *MIscore* and *MI³score* are as follows:

$$T - Score = \frac{f_{AB} - \frac{f_A f_B}{N}}{\sqrt{f_{AB}}} \tag{1}$$

$$MI - Score = log_2 \frac{f_{AB} N}{f_A f_B} \tag{2}$$

$$MI^3 - Score = log_2 \frac{f_{AB}^3 N}{f_A f_B} \tag{3}$$

The *Tscore*, *MIscore* and *MI³score* are applicable for processing diachronic corpora as well [11].

Collocations have been regarded as statistically similar words which can be extracted by using techniques for estimation the strength of association between co-occurring words. Recent developments improved that techniques with respect to various application areas.

Further, we shall present and analyze the search results for extracting conjunctive adverbs using the SE software and shall compare related results with respect to their semantic types.

4 Bulgarian Diachronic Mathematical Resource (BDMR)

The diachronic text corpora are designed to study predominantly how the grammar has changed over time. They typically use the annotation schemes which encode the grammatical relations by regarding them as relatively constant over certain period of time. The approach has been extensively used in diachronic corpora for languages like English, German, Spanish, etc. [6] allowing different types of semantic search. However, recent advances in IR offer the use of statistically-based approaches which allow comparison of different corpora with respect to several search criteria outlining both syntactic and lexical differences or changes without the use of syntactic annotations.

The Bulgarian Diachronic Mathematical Resource (BDMR) is the first collection of mathematical texts in Bulgarian language from the first half of XIX c. At the moment, it contains the first three original mathematical school books written in Bulgarian language.

The texts included are limited according to the related historical period and are representative both for the language of that period and for the level of mathematical knowledge and terminology at that time. The BDMR contains three school books: (i) 'Аритметика или наука числителна' (*Aritmetics or Science about Numbers*) by Chr. Pavlovich, published in 1833 in Belgrade [14]—consisting of almost 12 000 words, and (ii) 'Аритметическое руководство за наставление на болгарските юноши' (*Arithmetic Guide for Bulgarian Adolescents*) by N. Bozveli and Em. Vaskidovich, published in 1835 in Kraguevac [5]—consisting of almost 7 000 words, and (iii) 'Стихийная аритметика' (*Arithmetics in Verse*) by S. Radulov, published in 1843 in Smirna [15]—consisting of almost 23 000 words.

The BDMR uses electronic readable format of the books which are uploaded in their original graphical representation and without normalizing the phonetic alternations according to contemporary spelling rules. The resource was created to analyze the grammar and lexical features as well as to study the mathematical terminology of Bulgarian language for the related period. The BDMR was uploaded into the SE allowing the use of its incorporated options for storing, sampling, searching and filtering the texts according to different criteria. The resource is open for further development and enlargement.

5 Search Results

5.1 *Keyness Score*

To analyze the semantics of connectives used in the texts of BDMR, first we need to extract them using statistical search techniques [18]. For that, we have run several search experiments within every one of the parts of that electronic resource.

Single-word		Score	F	RefF
☐ перст	W	4,516.41	83	2
☐ день	W	4,454.58	90	86
☐ кольку	W	2,836.64	52	0
☐ быва	W	2,673.05	49	0
☐ тольку	W	2,563.98	47	0
☐ сумма	W	2,500.56	46	3
☐ остаток	W	2,343.08	43	1
☐ третій	W	2,236.79	41	0
☐ внетрешен	W	2,127.73	39	0
☐ Примѣр	W	2,127.73	39	0
☐ вторый	W	1,909.61	35	0
☐ став	W	1,827.59	85	1,296
☐ тамо	W	1,767.51	46	354
☑ сирѣчь	W	1,746.01	32	0
☐ знаменатель	W	1,746.01	32	0

Single-word		Score	F	RefF
☐ тѣхъ	W	4,703.90	64	13
☐ послѣ	W	4,104.87	55	0
☐ отъ	W	3,437.25	98	951
☐ тогда	W	2,591.89	36	31
☐ тако	W	2,307.72	41	275
☐ съ	W	2,037.35	106	2,431
☐ чрезъ	W	1,986.59	28	44
☐ подъ	W	1,896.44	27	53
☐ ока	W	1,754.06	33	341
☑ сирѣчь	W	1,717.16	23	0
☐ какъ	W	1,595.26	22	25
☐ гро	W	1,507.17	21	34
☐ мѣсяцы	W	1,493.31	20	0

Fig. 1 The *keyness score* top-down search results from BDMR (i) and BDMR (ii) for non-content words

For our search experiments, we use the techniques of corpora comparison. The main idea is to use the BDMR as a basic resource which to compare to arbitrary reference corpus consisting of texts from standard contemporary Bulgarian language. In that way, we use the *keyness score* measurement [8], so to extract the non-content words (generally prepositions and possibly conjunctive adverbials or conjunctions). We expect that they have to be specific both for the time period (the first half of XIX c.) and for the related domain (mathematics). The results from processing BDMR (i) and BDMR (ii) are presented at Fig. 1 and are similar.

They both include the word сирѣчь(*or*) which is considered as a specific for both the time period and for the mathematical domain because it has 32 hits for the BDMR (i), 23 hits for the BDMR (ii), and 0 hits for the reference corpus used. The results from the BDMR (iii) which are presented at Fig. 2 do not contain that word but contain another word послѣ (*after, later*), which is common for both BDMR (ii) and BDMR (iii) and has 0 hits for the reference corpus as well. That word also presents another possible connective.

The later results displayed for the BDMR (i) and BDMR (iii) are given at Fig. 3. They outline also another non-content word слѣдователно (*consequently, hence, therefore*) which has 21 hits for BDMR (i), 39 hits for the BDMR (iii), and 0 hits for the reference corpus as well. At the same time, the results present another word which is common for the BDMR (i), BDMR (ii) and BDMR (iii). It is the contraction of сирѣчь – сир.

However, the *keyness score* measurement do not give any additional information about the extracted keywords' syntactic distribution, so to evaluate their syntactic functions.

Single-word		Score	F	RefF
☐ отъ	W	4,598.43	334	951
☐ въ	W	4,127.67	375	1,401
☐ единицы	W	3,662.29	125	0
☐ изъ	W	3,602.48	125	14
☐ цифры	W	2,549.26	87	0
☐ дробь	W	2,022.03	69	0
☐ дѣлимо	W	1,934.16	66	0
☐ часть	W	1,816.55	65	41
☐ съ	W	1,773.08	235	2,431
☑ послѣ	W	1,758.42	60	0
☐ десетицы	W	1,729.13	59	0
☐ перво	W	1,673.11	60	43

Fig. 2 The *keyness score* top-down search results from BDMR (iii) for non-content words

				☑ сир	W	1,314.06	51	116
☐ ставы	W	1,418.82	26	0	☐ арш	W	1,244.25	43	11
☐ найда	W	1,380.93	52	889	☐ знаменатель	W	1,231.19	42	0
☐ тѣх	W	1,364.29	25	0	☐ дѣйство	W	1,231.19	42	0
☐ человѣк	W	1,309.76	24	0	☐ найдеме	W	1,228.28	42	2
☐ раздробленіе	W	1,309.76	24	0	☐ же	W	1,224.59	68	529
☐ онай	W	1,300.51	24	6	☐ подъ	W	1,158.39	42	53
☐ четвертый	W	1,255.23	23	0	☐ числител	W	1,143.35	44	108
☐ круг	W	1,162.11	22	28	☑ слѣдователно	W	1,143.32	39	0
☑ слѣдователно	W	1,146.16	21	0					

Fig. 3 The *keyness score* top-down search results from BDMR (i) and BDMR (iii) for non-content words (continuation)

5.2   *Concordances Generation*

For that, we use the approach already presented in [17] and used for cross-lingual mathematical terminology extraction. The main techniques applied are concordances generation and collocations generation. The concordances present all occurrences of a related keyword within all its quantitative contexts. Thus, we generate concordances for the keyword сирѣчь for both BDMR (i) and BDMR (ii) and for the contraction сир for BDMR (iii).

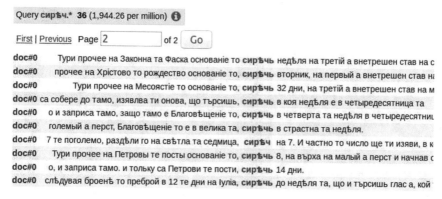

Fig. 4 The concordances of the keyword сирѣчь from BDMR (i)

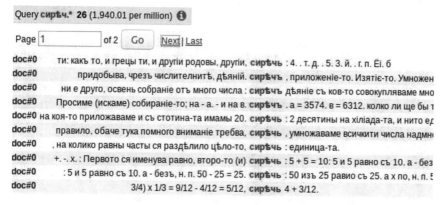

Fig. 5 The concordances of the keyword сирѣчь from BDMR (ii)

The results received are presented at Fig. 4, Fig. 5, and Fig. 6, respectively and display all related contexts that can be analyzed with respect to the syntactic variability and distribution of the keyword.

The context analysis shows that the word сирѣчь (and its contraction сир) which is an adverb is used to assign a syntactic relations of coordination (often replaceable by the coordinating conjunction или (*or*) or by тоест (*that is, namely*). It marks a relation of logical equivalence (linguistic synonymy) between the coordinating phrases or clauses and can be regarded as an adverbial logical transition maintaining the text coherence. Its semantics is to show similarity, to signal restatement, etc.

The results obtained for concordances generation of the keyword слѣдователно from BDMR (i) and BDMR (iii) are presented at Fig. 7 and Fig. 8, respectively and show the syntactic variability and distribution of that keyword. Also, the results for concordances generation of its contraction сл. from BDMR (iii) are presented at Fig. 9

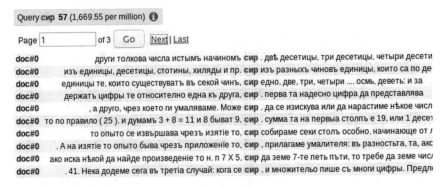

Fig. 6 The concordances of the contraction сир from BDMR (iii)

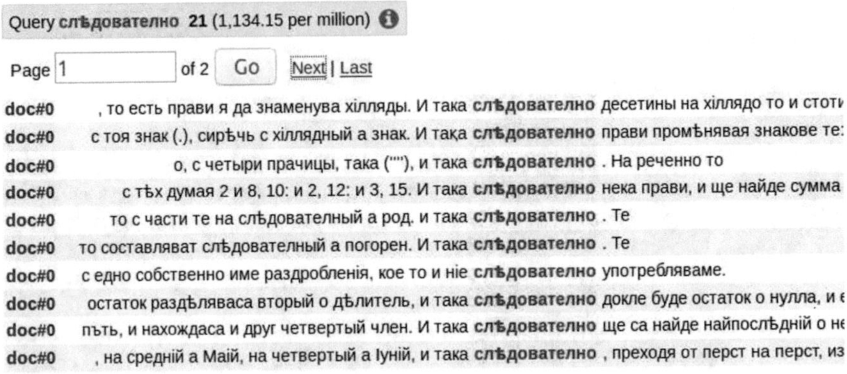

Query слѣдователно 21 (1,134.15 per million)

Page 1 of 2 Go Next | Last

doc#0 , то есть прави я да знаменува хілляды. И така **слѣдователно** десетины на хілляадо то и стоти

doc#0 с тоя знак (.), сирѣчь с хіллядный а знак. И така **слѣдователно** прави промѣнявая знакове те:

doc#0 о, с четыри прачицы, така (""), и така **слѣдователно** . На реченно то

doc#0 с тѣх думая 2 и 8, 10: и 2, 12: и 3, 15. И така **слѣдователно** нека прави, и ще найде сумма

doc#0 то с части те на слѣдователный а род. и така **слѣдователно** . Те

doc#0 то составляват слѣдователный а погорен. И така **слѣдователно** . Те

doc#0 с едно собственно име раздроблнія, кое то и ніе **слѣдователно** употребляваме.

doc#0 остаток раздѣлявааса вторый о дѣлител, и така **слѣдователно** докле буде остаток о нулла, и ɛ

doc#0 пъть, и нахождааса и друг четвертый член. И така **слѣдователно** ще са найде найпослѣдній о нɛ

doc#0 , на средній а Маій, на четвертый а Iуній, и така **слѣдователно** , преходя от перст на перст, из

Fig. 7 The concordances of the keyword слѣдователно from BDMR (i)

Query слѣдователно 42 (1,230.19 per million)

Page 1 of 3 Go Next | Last

doc#0 , става четыри: отъ четыри те, петь: и така **слѣдователно** до число то десеть. 10. А число то

doc#0 шестаго чина, петь единицы петого чина, и **слѣдователно** Но понеже това средство предста

doc#0 понятіе за нихъ ще пріимеме **слѣдователно** в тѣхните во особъ дѣйства. Прил

doc#0 умалителя 2 единицы, 5 десетицы и 4 стотини. **слѣдователно** 452 единицы. 30. Кога нѣкоя цифр

doc#0 ставатъ 13, а 13 - 5 = 8, пишемъ 8 подъ десетицы те. **слѣдователно** умалявамъ 5-те с единица. и така

doc#0 измѣненіе: защото може да се представи какъ **слѣдователно** 700000 + 30000 + 12000 + 400 + 1ǃ

doc#0 пріиматъ, цѣна десетижды поголѣма (15). и **слѣдователно** сичко то число се умножава съ 10.

doc#0 7 рѣдове, секой изъ които содержава 5 единицы и **слѣдователно** сумма та имъ се нахожда и така, с

doc#0 то 24 се изема, като извадихме три пъти 8-те, и **слѣдователно** число то 24 содержава 8-те точно

doc#0 въ дѣлимо то повече отъ 4, помалко же отъ 5, и **слѣдователно** цѣло то частно се нахожда между

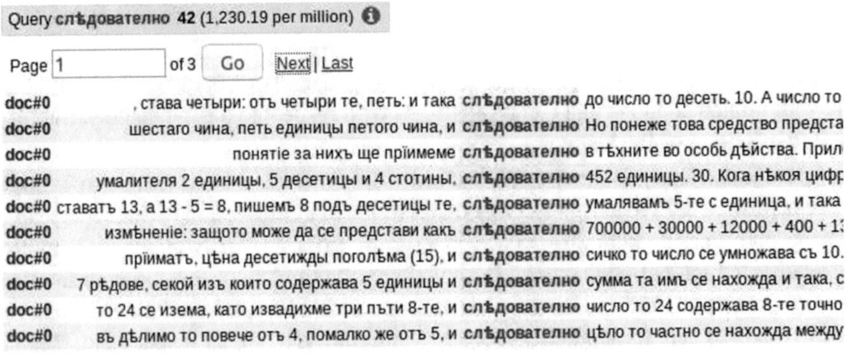

Fig. 8 The concordances of the keyword слѣдователно from BDMR (iii)

Query сл **15** (439.35 per million) ⓘ

doc#0	, и въ сумма та имъ, единицы те на третьо то, и така сл . Такова дѣйство новоначални те извършаватъ
doc#0	остатокъ, вторыа остатокъ на третіа, и т. сл . до гдѣ да найдеме остатокъ нулла: послѣдніо
doc#0	десеточастія, сотночастія, тысящночастія и сл . се земе единожды или многажды, дробьта, щ(
doc#0	та, сточастія та тысящночастія та и сл . изъ които се составляватъ десетичны те дроі
doc#0	сир. секоя цифра, кога проходи, едно, две, три и сл . положенніа на лѣво, да показува десеть, сто,
doc#0	на лѣво, да показува десеть, сто, хилядо пъти и сл . погорни единицы, и напротивъ, кога проходи
doc#0	, и напротивъ, кога проходи едно, две, три и сл . положенія на десно, да показува единицы
doc#0	, да показува единицы десеть, сто, хиляда пъти и сл . подолни. И понеже въ цѣли те числа прости т
doc#0	, явно е, че ако и въ перво то, второ, третьо и сл . положеніе послѣ единицы ти надѣсно напиш
doc#0	, ще значатъ единицы десеть, сто, хиляда и сл . пъти подолны отъ просты те единицы, или

Fig. 9 The concordances of the contraction сл. from BDMR (iii)

Collocation candidates

	Cooccurrence count	Candidate count	T-score	MI	MI3
P|N така	13	104	3.561	6.369	13.770
P|N и	17	426	3.966	4.721	12.896
P|N разрѣшавашь	5	11	2.228	8.231	12.875
P|N ,	23	1,178	4.424	3.690	12.737
P|N Найди	5	13	2.227	7.990	12.634
P|N направи	5	13	2.227	7.990	12.634
P|N а	12	280	3.341	4.824	11.994
P|N .	18	1,065	3.863	3.482	11.822

Fig. 10 The collocation candidates of the keyword слѣдователно from BDMR (i)

The results contain contexts which present interesting distribution and variability of the keyword, and show that the word слѣдователно (and its contraction сл.) is an adverb and assigns a syntactic relations of subordination (often introduced by the subordinating conjunction и така (и т.) (*so that, in order that*). Its syntactic function is to introduce an adverbial clause which modify the meaning of a previous clause (modifying the previous logical relation) and expressing the semantics of 'to introduce', 'to show causality', etc.

5.3 Collocations Generation

Thus, to analyze the syntactic variability of the word слѣдователно (and its contraction сл.) in more details, we need to use a search optimization that is aimed at further query profiling by generation of keyword's collocations. The related results for collocation candidates search of that keyword from the BDMR (i) are presented at Fig. 10. They outline a variability of the keyword's use in combination with a conjunctions и (*and*) and така (*thus*) which are very frequent collocations, however the combinations with a comma (,) are the most frequent (23 hits).

The results for collocation candidates search of the keyword слѣдователно (and its contraction сл.) from the BDMR (iii) are presented at Fig. 11 and Fig. 12,

Fig. 11 The collocation candidates of the keyword слѣдователно from BDMR (iii)

Collocation candidates

	Cooccurrence count	Candidate count	T-score	MI	MI3
P|N и	27	905	4.981	4.600	14.109
P|N ,	34	2,655	5.270	3.379	13.554
P|N .	17	1,634	3.635	3.080	11.255
P|N то	13	769	3.343	3.780	11.181
P|N какъ	4	36	1.977	6.496	10.496
P|N частно	4	77	1.952	5.400	9.400
P|N да	8	683	2.531	3.251	9.251
P|N 24	3	43	1.701	5.825	8.995

Fig. 12 The collocation candidates of the contraction сл. from BDMR (iii)

Collocation candidates

	Cooccurrence count	Candidate count	T-score	MI	MI3
P|N и	15	905	3.770	5.237	13.051
P|N сто	3	10	1.729	9.415	12.585
P|N .	15	1,634	3.687	4.385	12.198
P|N три	5	64	2.223	7.474	12.118
P|N ,	14	2,655	3.429	3.585	11.199
P|N две	3	56	1.717	6.929	10.099
P|N пъти	3	77	1.712	6.470	9.640
P|N се	3	417	1.626	4.033	7.203

respectively. They outline similar syntactic variability as the results from BDMR (i), and show that the keyword is used mostly in combination with the conjunction и (*and*) but the combinations with a comma (,) are most frequent (34 hits for слѣдователно and 14 hits for сл.).

However, a more detailed semantic analysis of the keyword's syntactic variability shows that the keyword is used with two different meanings (in a combinations with different conjunctions). In some contexts the word слѣдователно (especially in combination with и така (и т.)) is used instead of последователно. However, the majority of contexts (the use in a combination with the conjunction и or single use) outline its functioning as a logical connective.

5.4 Semantic Clusters

Further work on the search optimization is done, so to outline the syntactic functionality of the keyword слѣдователно in more details by revealing its hidden semantic relations. The search technique that is used is indebted to the use of *fuzzy sets* ideas [10] and is based on the comparison of collocations of a related keywords.

In that way, it is possible to perform a search, so to compare all possible collocations of a related keyword tracking whether some of that collocations belong also to another keyword. The received search results show the hidden semantic relations between the keyword and its related keywords (sharing common collocations) and form a semantic cluster of words which are semantically related to a keyword (semantic synonyms).

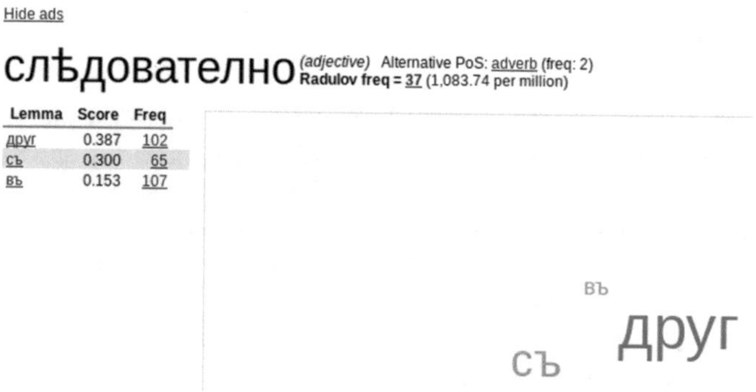

слѣдователно *(adjective)* Alternative PoS: <u>adverb</u> (freq: 2)
Radulov freq = <u>37</u> (1,083.74 per million)

Lemma	Score	Freq
<u>друг</u>	0.387	<u>102</u>
<u>съ</u>	0.300	<u>65</u>
<u>въ</u>	0.153	<u>107</u>

въ

съ   **друг**

Fig. 13 The semantic cluster results for the keyword слѣдователно from BDMR (iii)

слѣдователно *(adjective)* Alternative PoS: <u>adverb</u> (freq: 2)
Radulov freq = <u>37</u> (1,083.74 per million)

a_modifier			pp_в		
		35.14			2.70
кубо	1	11.19	тѣхните	1	13.99
слѣдователно кубо			слѣдователно в тѣхните		
удобоупотребителны	1	11.19			
слѣдователно удобоупотребителны					
тѣхните	1	11.19			
слѣдователно в тѣхните					
корень	1	11.09			
слѣдователно найденныо корень					
сумма	1	11.09			
слѣдователно сумма					
тръсиме	1	11.09			
слѣдователно тръсиме					
5-те	1	11			
слѣдователно умалявамъ					
цѣло	1	11			
слѣдователно цѣло					
корень	1	10.35			
слѣдователно цѣлыо корень					

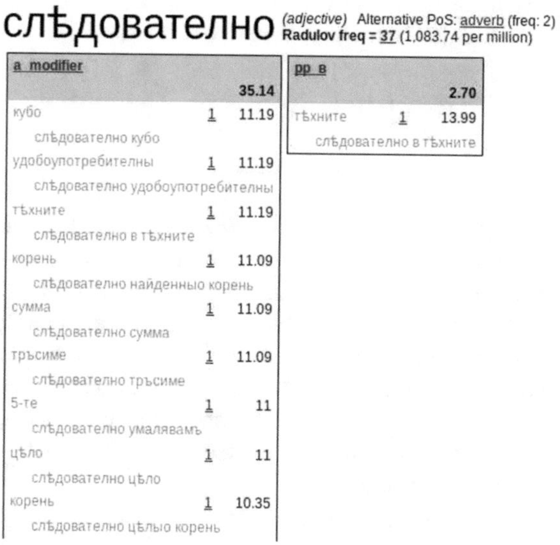

Fig. 14 The grammatical and semantic relations of the keyword слѣдователно from BDMR (iii)

The results from that type of search for the keyword слѣдователно from BDMR (iii) are presented at Fig. 13 and outline that the prepositions в (*in*) and с (*with*) (which are also connectives) are semantically related to that keyword together with the word друг (*another*). The results confirm the previous claim that the syntactic function of the keyword слѣдователно is a logical connective.

5.5 *Grammatical Relations*

However, the semantic relations of the keyword слѣдователно which define its lexical meaning can be outlined by presenting both its grammatical and its semantic relations containing all related syntactic functions. That is done by generating the keyword's sketch profile. The search results of the keyword слѣдователно from the BDMR (iii) are presented at Fig. 14 and outline the grammatical relations like 'modifier', 'pp–в' as well as the semantic relations like 'and/or' which improve the previously received results—confirming that the word is a modifier and is semantically related to the preposition в (*in*).

Thus, the word слѣдователно functions as an adverbial modifier, i. e. is a logical connective. Also, it can be regarded as a subordinating transition which semantics is to introduce an item in a series (sequence), to introduce an example, to show causality (purpose or reason), to introduce a summary or conclusion.

6 Conclusion

We presented an IR approach which uses statistically-based techniques to search electronic collection of diachronic mathematical texts containing the first three school books written in Bulgarian language during the first half of XIX c. The aim was to extract and study the syntactic phenomena, i. e. the logical connectives which are specific both to that period of time and to the domain of mathematics. We have extracted conjunctive adverbs by using keyness score measurement and have used concordances search, collocations generation, and semantic clusters, so to optimize the search. Later, we have analyzed the related results both with respect to their syntactic distribution and variability.

The obtained results outlined the words сирѣчь and слѣдователно which were analyzed from the point of view of text coherence, and were regarded as logical connectives (transitions). The former was evaluated as a coordinating transition whereas the later was evaluated as a subordinating transition, and the related semantics was defined for both, respectively.

Finally, from the point of view of diachronic syntax, the adverbial сирѣчь is not used any more in contemporary Bulgarian language and is regarded as stylistically archaic, whereas the adverbial слѣдователно is very actively used in contemporary academic writing, and not only in the domain of mathematics. Obviously, this is one of the first use of that adverb in Bulgarian language (with the above described syntactic functions and semantics), since it defines the semantic relations between the connected phrases or clauses explicitly and maintains their logical interconnections.

References

1. Baroni, M., Evert, S.: Statistical methods for corpus exploitation. In: Lüdeling, K., Kytö, M. (eds.) Corpus Linguistics: An International Handbook, vol. 2, pp. 777–803. Mouton De Gruyter, Berlin, New York (2008)
2. Baroni, M., Lenci, A.: Distributional memory: a general framework for corpus-based semantics. Comput. Linguist. **36**(4), 673–721 (2010)
3. Beilin, H., Barbara, L.: A study of the development of logical and linguistics connectives: linguistics data. In: Studies in the Cognitive Basis of Language Development, pp. 76–120 (1975)
4. Bloom, L., et al.: Complex sentences: acquisition of syntactic connectives and the semantic relations they encode. J. Child Lang. **7**(02), 235–261 (1980)
5. Бозвели, Н., Васкидович, Ем.: Аритметическое руководство за наставление на болгарските юноши. Княжеско-Сербска типография, Крагуевац (1835) (in Bulgarian language)
6. Davies, M., Chapman, D.: The effect of representativeness and size in historical corpora: an empirical study of changes in lexical frequency. In: Chapman, D., Moore, C., Wilcox, M. (eds.) Studies in the History of the English Language VII: Generalizing vs. Particularizing Methodologies in Historical Linguistic Analysis, pp. 131–150. De Gruyter/Mouton, Berlin (2016)
7. Falmagne, R. J.: Language and the acquisition of logical knowledge. In: Reasoning, Necessity, and Logic: Developmental Perspectives, pp. 111–131 (1990)
8. Gabrielatos, C.: Keyness analysis: nature, metrics and techniques. In: Taylor, C., Marchi, A. (eds.) Corpus Approaches to Discourse: A Critical Review, pp. 225–258. Routledge, Oxford (2018)
9. Humberstone, L.: The Connectives. MIT Press, Cambridge, MA (2011)
10. Ismaili, S., Fidanova, S.: Applications of intuitionistic Fuzzy sets on agent based modeling. Proc. Bulg. Acad. Sci. **71**(6), 812–819 (2018)
11. Killgarriff, A., Husak, M., Woodrow, R.: The Sketch Engine as infrastructure for historical corpora. In: Jancsary, J. (ed.) Empirical Methods in Natural Language Processing, Proceedings of the Conference on Natural Language Processing 2012, pp. 351–356 (2012)
12. Killgarriff, A., et al.: The Sketch Engine: ten years on. Lexicography **1**, 17–36 (2014)
13. Miltsakaki, E., et al.: Annotating discourse connectives and their arguments. In: Proceedings of the HLT/NAACL Workshop on Frontiers in Corpus Annotation (2004)
14. Павлович, Хр.: Аритметика или наука числителна. Княжеско-Сербска типография, Белград (1833) (in Bulgarian language)
15. Радулов, С.: Стихйная аритметика во употребления болгарскихъ училищъ. Типография А. Дамянова, Смирна (1843) (in Bulgarian language)
16. Rappaport, B.: Using the elements of rhythm, flow, and tone to create a more effective and persuasive acoustic experience in legal writing. J. Leg. Writ. Inst. **16**(1), 65–116 (2010)
17. Stoykova, V., Stankovic, R.: Using query expansion for cross-lingual mathematical terminology extraction. In: Silhavy, R. (ed.) Advances in Intelligent Systems and Computing (AISC), vol. 764, pp. 154–164. Springer (2018)
18. Stoykova, V.: The statistic analysis of conjunctive adverbs used in the first Bulgarian school books in mathematics (from the first half of XIX c). In: Nikolov, G., Kolkovska, N., Georgiev, K, (eds.) 9th International Conference on Numerical Methods and Applications, NMA 2018, Borovets, Bulgaria. LNCS, vol. 11189. Springer, pp. 150–157 (2019)

An Improved "Walk on Equations" Monte Carlo Algorithm for Linear Algebraic Systems

Venelin Todorov, Nikolay Ikonomov, Stoyan Apostolov, Ivan Dimov, Rayna Georgieva and Yuri Dimitrov

Abstract A new Monte Carlo algorithm for solving systems of Linear Algebraic (LA) equations is presented and studied. The algorithm is based on the "Walk on Equations" Monte Carlo method recently developed by Dimov et al. (Appl Math Model 39:4494–4510, [12]). The algorithm is improved by choosing the appropriate values for the relaxation parameters which leads to dramatic reduction in time and lower relative errors for a given number of iterations. Numerical tests are performed for examples with matrices of different size and on a system coming from a finite element approximation of a problem describing a beam structure in constructive mechanics.

V. Todorov (✉) · N. Ikonomov · Y. Dimitrov
Department of Information Modeling, Institute of Mathematics and Informatics,
Bulgarian Academy of Sciences, Acad. Georgi Bonchev Str., Block 8,
1113 Sofia, Bulgaria
e-mail: vtodorov@math.bas.bg; venelin@parallel.bas.bg

N. Ikonomov
e-mail: nikonomov@math.bas.bg

Y. Dimitrov
e-mail: yuri.dimitrov@ltu.bg

V. Todorov · I. Dimov · R. Georgieva
Department of Parallel Algorithms, Institute of Information and Communication Technologies,
Bulgarian Academy of Sciences, Acad. G. Bonchev Str., Block 25 A, 1113 Sofia, Bulgaria
e-mail: ivdimov@bas.bg

R. Georgieva
e-mail: rayna@parallel.bas.bg

S. Apostolov
Faculty of Mathematics and Informatics, Sofia University, Sofia, Bulgaria
e-mail: stoyanrapostolov@gmail.com

Y. Dimitrov
Department of Mathematics and Physics, University of Forestry, Sofia, Bulgaria

© Springer Nature Switzerland AG 2020
S. Fidanova (ed.), *Recent Advances in Computational Optimization*,
Studies in Computational Intelligence 838,
https://doi.org/10.1007/978-3-030-22723-4_14

1 Introduction

Many scientific and engineering applications are based on the problems of solving systems of linear algebraic (LA) equations [4, 5, 18, 21, 22, 47]. For some applications it is also important to compute directly the inner product of a given vector and the solution vector of a linear algebraic system [1, 10]. It is also very important to have relatively *cheap* algorithms for matrix inversion [9, 14]. The computation time for very large problems, or for finding solutions in real-time, can be prohibitive and this prevents the use of many established algorithms. Monte Carlo methods are methods of approximation of the solution to problems of computational mathematics, by using random processes for each such problem, with the parameters of the process equal to the solution of the problem [2, 8, 31]. The method can guarantee that the error of Monte Carlo approximation is smaller than a given value with a certain probability [19, 48].

In Monte Carlo (MC) numerical algorithms we construct a Markov process and prove that the mathematical expectation of the process is equal to the unknown solution of the problem [8]. A Markov process is a stochastic process that has the Markov property. Often, the term Markov chain is used to mean a discrete-time Markov process [11].

Definition 1 A finite discrete Markov chain T_k is defined as a finite set of states $\{\alpha_1, \alpha_2, \ldots, \alpha_k\}$.

Definition 2 A state is called absorbing if the chain terminates in this state with probability one.

Iterative Monte Carlo algorithms can be defined as terminated Markov chains [7, 45]:

$$T = \{\alpha_{t_0} \to \alpha_{t_1} \to \alpha_{t_2} \ldots \alpha_{t_k}\}, \tag{1}$$

where α_{t_q}, $q = 1, \ldots, i$ is one of the absorbing states.

An important advantage of the MC algorithms is that often a slight modification of the algorithm for solving systems of LA equations allows to solve other linear algebra problems like matrix inversion and computing the extremal eigenvalues. In this work algorithms for solving systems of LA equations are presented. It is also shown what some modifications are needed to be able to apply these algorithms for a wider class of linear algebra problems. In [6] a MC method for solving a class of linear problems is presented. It is proved that the method is closed to optimality in sense that for a special choice of the right-hand side the requirements of the Kahn theorem are fulfilled and the method is optimal. This method is called almost optimal method (MAO). The MAO method was exploit by C. Jeng and K. Tan in [32] to build on the top of MAO the relaxed MC method for systems of LA problems. The idea of the approach used in [32] is to use a relaxation parameter γ like in deterministic Gauss-Seidel algorithm, or Jacobi overrelaxation algorithm. In [45] an unbiased estimation of the solution of the system of linear algebraic equations is presented. In [23] a MC algorithm for matrix inversion is proposed and studied. The algorithm is based on the solution of simultaneous linear equations. There are MC algorithms for computing components of the solution vector; evaluating linear functionals of the solution; matrix inversion, and computing the extremal eigenvalues. An overview of all these algorithms is given in [8]. The well-known Power method is described in [25]. In [17] the robustness and applicability of the Almost Optimal Monte Carlo algorithm for solving a class of linear algebra problems based on bilinear form of matrix powers is analysed. Several authors have presented works on MC algorithms for LA problems and on the estimation of computational complexity [1, 9–11, 14, 19]. The Resolvent Power Monte Carlo method is described in several recent papers [10, 15, 16]. The approach is a discrete analogue of the resolvent analytical continuation method used in the functional analysis [34]. The computational complexity can be decreased by the use of acceleration parameter based on the resolvent representation or to apply a variance reduction technique [5] in order to get the required approximation of the solution with a smaller number of operations.

The rest of the paper is organized as follows. In Sect. 2, we formulate the problem of solving linear systems using Monte Carlo methods based on Markov chains with or without absorbing states. Also some analysis of the variance has been done. In Sect. 3, we describe the improved "Walk on Equations" (WE) Monte Carlo method for solving linear systems. We also explain how to accelerate the convergence of the algorithms thanks to the sequential Monte Carlo approach and how to optimize the relaxation parameter which leads to the balancing of the iteration matrix. In Sect. 4, we make numerical comparisons between our method, the original method, refined iterative Monte Carlo method and the preconditioned conjugant gradient for the approximation of all components of large linear systems. In the end some concluding remarks for the benefits of the improved method are given.

2 Description of the Monte Carlo Algorithm for Linear Systems

By A and B we denote matrices of size $n \times n$, i.e., $A, B \in \mathbb{R}^{n \times n}$. We use the following presentation of matrices:

$$A = \{a_{ij}\}_{i,j=1}^{n} = (a_1, \ldots, a_i, \ldots, a_n)^t,$$

where $a_i = (a_{i1}, \ldots, a_{in})$, $i = 1, \ldots, n$ and the symbol t means *transposition*.
The following norms of vectors:

$$\| b \| = \| b \|_1 = \sum_{i=1}^{n} |b_i|, \quad \| a_i \| = \| a_i \|_1 = \sum_{j=1}^{n} |a_{ij}|$$

and matrices

$$\| A \|_1 = \max_j \sum_{i=1}^{n} |a_{ij}|, \quad \| A \|_\infty = \max_i \sum_{j=1}^{n} |a_{ij}|$$

are used, where $b \in \mathbb{R}^n$. We consider a system of LA equations

$$Bx = f, \tag{2}$$

where $B = \{b_{ij}\}_{i,j=1}^{n} \in \mathbb{R}^{n \times n}$ is a given matrix; $f = (f_1, \ldots, f_n)^t \in \mathbb{R}^{n \times 1}$ and $v = (v_1, \ldots, v_n) \in \mathbb{R}^{1 \times n}$ are given vectors.

We deal with the matrix $A = \{a_{ij}\}_{ij=1}^{n}$, such that $A = I - DB$, where D is a diagonal matrix $D = diag(d_1, \ldots, d_n)$ and $d_i = \frac{\gamma}{b_{ii}}$, $i = 1, \ldots, n$, and $\gamma \in (0, 1]$ is a parameter that can be used to accelerate the convergence. The system (2) can be presented in the form of equation

$$x = Ax + b, \tag{3}$$

where $b = Df$. Let us suppose that the matrix B is diagonally dominant. Obviously, if B is a diagonally dominant matrix, then the elements of the matrix A must satisfy the following condition:

$$\sum_{j=1}^{n} |a_{ij}| \leq 1, \quad i = 1, \ldots, n.$$

It is possible to choose a non-singular matrix [3, 25] $M \in \mathbb{R}^{n \times n}$ such that $MB = I - A$, where $I \in \mathbb{R}^{n \times n}$ is the identity matrix and $Mf = b, b \in \mathbb{R}^{n \times 1}$. Consider the system:

$$x = Ax + b. \tag{4}$$

It is assumed that

(i) $\begin{cases} 1. \text{ The matrices } M \text{ and } A \text{ are both non-singular;} \\ 2. \ |\lambda(A)| < 1 \ \text{ for all eigenvalues } \lambda(A) \text{ of } A, \end{cases}$

i.e. for all values $\lambda(A)$ of the matrix A, for which $Ay = \lambda(A)y$. If the conditions (i) are satisfied, then A *stationary linear iterative algorithm* [8, 11] can be used:

$$x_k = Ax_{k-1} + b, \quad k = 1, 2, \ldots \tag{5}$$

and the solution x can be presented in a form of a Neumann series

$$x = \sum_{k=0}^{\infty} A^k b = b + Ab + A^2 b + A^3 b + \cdots \tag{6}$$

The *stationary linear iterative Monte Carlo algorithm* is based on (6) [32]. As a result, the convergence of the Monte Carlo algorithm depends on the truncation error of the series (5) (see, [8]). Analysis of convergence conditions in Monte Carlo algorithms for linear systems based on stationary linear iterative methods is presented in some recent papers [16, 33, 49]. We are interested to evaluate the linear form $W(x)$ of the solution x of the system (4), i.e.,

$$W(x) \equiv (w, x) = \sum_{i=1}^{n} w_i x_i, \qquad (7)$$

where $w \in \mathbb{R}^{n \times 1}$. We shall define a random variable $X[w]$, which expectation is equal to the above defined linear form, i.e., $EX[w] = W(x)$ using a discrete Markov process with a finite set of states. Then the computational problem is to calculate repeated realizations of $X[w]$ and of combining them into an appropriate statistical estimator of $W(x)$. It is clear that if we choose the special case of linear form $W(x)$ according to vector $w = e_i = (0, 0, \ldots, \underbrace{1}_{i}, 0, \ldots, 0)$, where the one is in the i-th place we get the i-th component of the solution x_i. The original algorithm [12] is based on counting the number of visits of each equation we compute all the components of the solution x.

A slight modification of $X[w]$ allows to compute an approximation of the inverse matrix [12]. Indeed, to compute the inverse matrix $G = B^{-1}$, of $B \in \mathbb{R}^{n \times n}$ one needs to assume that the matrix B is non-singular, and $||\lambda(B)| - 1| < 1$ for all eigenvalues $\lambda(B)$ of B and the *iterative* matrix is $A = I - B$. Then, the inverse matrix can be presented as $G = \sum_{i=0}^{\infty} A^i$. According to [12], $E\{\sum_{i|k_i=r'} Q_i\} = g_{rr'}$, where $(i|k_i = r')$ means a summation only for weights Q_i for which $k_i = r'$ and $G = \{g_{rr'}\}_{r,r'=1}^{n}$. Another modification of $X[w]$ allows to compute first $k \leq n$ eigenvalues of real symmetric matrices, since in this case the eigenvalues are real numbers (see, [12, 17]). The proposed Monte Carlo approach for solving systems of LA equations can easily be modified for a wider class of linear algebra problems [13, 15].

Consider an initial density vector $p = \{p_i\}_{i=1}^{n} \in \mathbb{R}^n$, such that $p_i \geq 0, i = 1, \ldots, n$ and $\sum_{i=1}^{n} p_i = 1$. Consider also a transition density matrix $P = \{p_{ij}\}_{i,j=1}^{n} \in \mathbb{R}^{n \times n}$, such that $p_{ij} \geq 0$, $i, j = 1, \ldots, n$ and $\sum_{j=1}^{n} p_{ij} = 1$, for any $i = 1, \ldots, n$. Define sets of *permissible* densities $\mathcal{P}_b$ and $\mathcal{P}_A$.

Definition 3 The initial density vector $p = \{p_i\}_{i=1}^{n}$ is called permissible to the vector $v = \{v_i\}_{i=1}^{n} \in \mathbb{R}^n$, i.e. $p \in \mathcal{P}_b$, if

$$\begin{cases} p_{\alpha_s} > 0 & \text{when } v_{\alpha_s} \neq 0 \\ p_{\alpha_s} = 0 & \text{when } v_{\alpha_s} = 0. \end{cases} \qquad (8)$$

Similarly, the transition density matrix $P = \{p_{ij}\}_{i,j=1}^{n}$ is called *permissible* to the matrix $A = \{a_{ij}\}_{i,j=1}^{n}$, i.e. $P \in \mathcal{P}_A$, if

$$\begin{cases} p_{\alpha_{s-1},\alpha_s} > 0 & \text{when } a_{\alpha_{s-1},\alpha_s} \neq 0 \\ p_{\alpha_{s-1},\alpha_s} = 0 & \text{when } a_{\alpha_{s-1},\alpha_s} = 0. \end{cases} \tag{9}$$

We will be dealing with *permissible* densities [8] $\mathcal{P}_b$ and $\mathcal{P}_A$. It is obvious that in such a way the random trajectories constructed to solve the problems under consideration never visit zero elements of the matrix. Such an approach decreases the computational complexity of the algorithms. It is also very convenient when large sparse matrices need to be treated.

One possible permissible density can be chosen in the following way:

$$p = \{p_\alpha\}_{\alpha=1}^n \in \mathcal{P}_b, \quad p_\alpha = \frac{|v_\alpha|}{\| v \|};$$

$$P = \{p_{\alpha\beta}\}_{\alpha,\beta=1}^n \in \mathcal{P}_A, \quad p_{\alpha\beta} = \frac{|a_{\alpha\beta}|}{\| a_\alpha \|}, \alpha = 1, \dots, n. \tag{10}$$

Such a choice of the initial density vector and the transition density matrix leads to an *Almost Optimal Monte Carlo* (MAO) algorithm (see [6, 8, 11, 17]). The initial density vector $p = \{p_\alpha\}_{\alpha=1}^n$ is called *almost optimal initial density vector* and the transition density matrix $P = \{p_{\alpha\beta}\}_{\alpha,\beta=1}^n$ is called *almost optimal density matrix* [6]. Such density distributions lead to almost optimal algorithms in the sense that for a class of matrices A and vectors h such a choice coincides with optimal weighted algorithms defined in [20] and studied in [43] (for more details see [6]). In particular, the MAO algorithm becomes optimal (with a minimal variance) if all the elements of the right-hand side vector of the linear system are equal (see, [6]). The reason to use MAO instead of Uniform Monte Carlo is that MAO normally gives much smaller variances. On the other hand, the truly optimal weighted algorithms are very time consuming, since to define the optimal densities one needs to solve an additional integral equation with a quadratic kernel. This procedure makes the optimal algorithms very expensive.

Consider a real linear system of the form $x = Ax + b$ where the matrix A of size n is such that the convergence radius $\varrho(A) < 1$, its coefficients $a_{i,j}$ are real numbers and

$$\sum_{j=1}^n |a_{i,j}| \leq 1, \forall 1 \leq i \leq n.$$

We now define a Markov chain T_k with $n + 1$ states $\alpha_1, \dots, \alpha_n, n + 1$, such that

$$P(\alpha_{k+1} = j | \alpha_k = i) = |a_{i,j}|$$

if $i \neq n + 1$ and

$$P(\alpha_{k+1} = n + 1 | \alpha_k = n + 1) = 1.$$

We also define a vector c such that $c(i) = b(i)$ if $1 \leq i \leq n$ and $c(n + 1) = 0$. Denote by $\tau = (\alpha_0, \alpha_1, \dots, \alpha_k, n + 1)$ a random trajectory that starts at the initial state

$\alpha_0 < n + 1$ and passes through $(\alpha_1, \ldots, \alpha_k)$ until the absorbing state $\alpha_{k+1} = n + 1$. The probability to follow the trajectory τ is $P(\tau) = p_{\alpha_0} p_{\alpha_0 \alpha_1}, \cdots p_{\alpha_{k-1,k} \alpha_k} p_{\alpha_k}$. We use the MAO algorithm [3, 6, 25, 48] for the initial density vector $p = \{p_\alpha\}_{\alpha=1}^n$ and for the transition density matrix $P = \{p_{\alpha\beta}\}_{\alpha,\beta=1}^n$, as well. The weights Q_α are defined:

$$Q_m = Q_{m-1} \frac{a_{\alpha_{m-1}, \alpha_m}}{p_{\alpha_{m-1}, \alpha_m}}, \quad m = 1, \ldots, k, \quad Q_0 = \frac{c_{\alpha_0}}{p_{\alpha_0}}. \tag{11}$$

The estimator $X_\alpha(\tau)$ can be presented as

$$X_\alpha(\tau) = c_\alpha + Q_k \frac{a_{\alpha_k \alpha}}{p_{\alpha_k}}, \quad \alpha = 1, \ldots \tag{12}$$

taken with a probability $P(\tau) = p_{\alpha_0} p_{\alpha_0 \alpha_1}, \cdots p_{\alpha_{k-1,k} \alpha_k} p_{\alpha_k}$.

For the convergence of the process we use the following theorem [12].

Theorem 1 *The random variable $X_\alpha(\tau)$ is an unbiased estimator of x_α, i.e.*

$$E\{X_\alpha(\tau)\} = x_\alpha. \tag{13}$$

A probabilistic representation for complex-valued matrices is given in [12]. If we want to solve a linear system of the form $u = Au + b$, where the square matrix $A \in \mathbf{C}^{n \times n}$ with complex-valued entrances is of size n. Assume also that $\varrho(A) < 1$, its coefficients $a_{i,j}$ verify $\sum_{j=1}^n |a_{i,j}| \le 1$, for any i, where $1 \le i \le n$. We also define a Markov chain T_k with $n + 1$ states $\alpha_1, \ldots, \alpha_n, n + 1$, such that

$$P(\alpha_{k+1} = j | \alpha_k = i) = |a_{i,j}|$$

if $i \ne n + 1$ and

$$P(\alpha_{k+1} = n + 1 | \alpha_k = n + 1) = 1.$$

We also define a vector c such that $c(i) = b(i)$ if $1 \le i \le n$ and $c(n + 1) = 0$. Denote by $\tau = (\alpha_0, \alpha_1, \ldots, \alpha_k, n + 1)$ a random trajectory that starts at the initial state $\alpha_0 < n + 1$ and passes through $(\alpha_1, \ldots, \alpha_k)$ until the absorbing state $\alpha_{k+1} = n + 1$. The probability to follow the trajectory τ is $P(\tau) = p_{\alpha_0} p_{\alpha_0 \alpha_1}, \cdots p_{\alpha_{k-1,k} \alpha_k} p_{\alpha_k}$.

Now we define a random variable S_{α_k} such that

$$S_{\alpha_0} = 1, \quad S_{\alpha_k} = S_{\alpha_{k-1}} \exp\{i \arg(a_{\alpha_{k-1} \alpha_k})\},$$

where $\arg(x + iy) \equiv \tan^{-1}\left(\frac{y}{x}\right)$ is the complex argument. The above expression reduces to

$$S_{\alpha_0} = 1, \quad S_{\alpha_k} = S_{\alpha_{k-1}} sign(a_{\alpha_{k-1}, \alpha_k})$$

in the real case. Then, we have the following probabilistic representation:

$$x_\alpha = E\left\{c_\alpha + S_{\alpha_k}\frac{a_{\alpha_k\alpha}}{p_{\alpha_k}}\right\}, \quad \alpha = 1, \ldots, n. \tag{14}$$

Consider the variance of the random variable $X_\alpha(\tau)$ for evaluation the linear form for the solution $W(x)$. We use the following notations: $\overline{A} = \{|a_{ij}|\}_{i,j=1}^n, \widehat{c} = \{c_i^2\}_{i=1}^{n+1}$. The special choice of almost optimal probability densities leads to the Markov chain:

$$c_{\alpha_0} \rightarrow a_{\alpha_0\alpha_1} \rightarrow \cdots \rightarrow a_{\alpha_{k-1}\alpha_k}. \tag{15}$$

For this finite chain we have that

$$A_c^k = c_{\alpha_0}\prod_{s=1}^k a_{\alpha_{s-1}\alpha_s}, \tag{16}$$

where $c \in \mathbb{R}^{n\times 1}$ and $c(i) = b(i)$ if $1 \le i \le n$ and $c(n+1) = 0$. The variance of the random variable $X_\alpha^k(\tau)$ is defined as [12]

$$X_\alpha^k(\tau) = \frac{c_{\alpha_0}}{p_{\alpha_0}}\frac{a_{\alpha_0\alpha_1}}{p_{\alpha_0\alpha_1}}\frac{a_{\alpha_1\alpha_2}}{p_{\alpha_1\alpha_2}}\cdots\frac{a_{\alpha_{k-1}\alpha_k}}{p_{\alpha_{k-1}\alpha_k}}\frac{c_{\alpha_k}}{p_{\alpha_k}} = \frac{A_c^k c_{\alpha_k}}{P^k(\tau)}. \tag{17}$$

The variance of the random variable $X_\alpha^k(\tau)$ is very important for the quality of the algorithm. Smaller variance $Var\{X_\alpha^k(\tau)\}$ leads to better convergence of the stochastic algorithm. It is proven that [12]:

Theorem 2
$$Var\{X_\alpha^k(\tau)\} = \frac{c_{\alpha_0}}{p_{\alpha_0}p_\alpha}(\overline{A}_c^k\widehat{c})_\alpha - (A_c^k c)_\alpha^2. \tag{18}$$

The next theorem [12] is the key point in optimizing the basic Monte Carlo algorithm "Walk on Equations".

Theorem 3 *Consider a perfectly balanced stochastic matrix*

$$A = \begin{pmatrix} \frac{1}{n} & \cdots & \frac{1}{n} \\ & \vdots & \\ \frac{1}{n} & \cdots & \frac{1}{n} \end{pmatrix}$$

and the vector $c = (1, \ldots, 1)^t$. *Then MC algorithm defined by trajectory probability* $P^k(\tau)$ *is a zero-variance MC algorithm.*

Proof It is sufficient to show that the variance $Var\{X_\alpha^k(\tau)\}$ is zero. Obviously,

$$Ac = \begin{pmatrix} \frac{1}{n} & \cdots & \frac{1}{n} \\ & \vdots & \\ \frac{1}{n} & \cdots & \frac{1}{n} \end{pmatrix}\begin{pmatrix} 1 \\ \vdots \\ 1 \end{pmatrix} = \begin{pmatrix} 1 \\ \vdots \\ 1 \end{pmatrix},$$

and also,

$$A^k c = \begin{pmatrix} 1 \\ \vdots \\ 1 \end{pmatrix}.$$

Since

$$\bar{A} = \begin{pmatrix} \left|\frac{1}{n}\right| & \cdots & \left|\frac{1}{n}\right| \\ \vdots & & \\ \left|\frac{1}{n}\right| & \cdots & \left|\frac{1}{n}\right| \end{pmatrix} = A$$

and $\hat{c} = \{c_i^2\}_{i=1}^n = c$, we have: $\frac{|c_{\alpha_0}|}{p_{\alpha_0} p_\alpha} \times (\bar{A}_c^k \hat{c})_\alpha = 1$. Thus, we proved that $Var\{X_\alpha^k(\tau)\} = 0$ ◇

Let us mention that solving systems with perfectly balanced stochastic matrices does not make sense because this is not a realistic case for practical computations and the system is overdetermined. Nevertheless, it is possible to consider matrix-vector iterations $A^k c$ since they are the basics for the Neumann series that approximate the solution of systems of linear algebraic equations. The idea of this consideration was to demonstrate that, the closer the matrix A is to the perfectly balanced matrix, the smaller is the probability error of the algorithm.

3 A New Monte Carlo Algorithm for LA Systems

We describe on a very simple example how we obtain the probabilistic representation. The example is taken from [12] and it is given to show the main idea of the original WE Monte Carlo algorithm. We will assume that:

$$a_{ij} \geq 0, \ i, j = 1, \ldots, n, \ \sum_{j=1}^n a_{ij} \leq 1. \tag{19}$$

We consider the system of two equations:

$$x_1 = \frac{1}{2}x_1 + \frac{1}{4}x_2 + 1, \ x_2 = \frac{1}{3}x_1 + \frac{1}{3}x_2 + 2 \tag{20}$$

with unknowns x_1 and x_2. We have obviously $x_1 = 1 + E(X)$ where

$$P(X = x_1) = \frac{1}{2}, P(X = x_2) = \frac{1}{4}, P(X = 0) = \frac{1}{4} \tag{21}$$

and $x_2 = 2 + E(Y)$ where

$$P(Y = x_1) = \frac{1}{3}, \, P(Y = x_2) = \frac{1}{3}, \, P(Y = 0) = \frac{1}{3}.$$

Only one sample to approximate the expectations will be used and the score initialized by zero along a random walk on the two equations will be computed. First to approximate x_1 we add 1 to the score of our walk and then either we stop with probability $\frac{1}{4}$ because we know the value of X which is zero or either we need again to approximate $x(1)$ or $u(2)$ with probability $\frac{1}{2}$ or $\frac{1}{4}$. If the walk continues and that we need to approximate $x(2)$, we add 2 to the score of our walk, we stop with probability $\frac{1}{3}$ or either we need again to approximate x_1 or $x(2)$ with probability $\frac{1}{3}$. The walk continues until one of the random variables X or Y takes the value zero and the score is incremented along the walk as previously. The score is an unbiased estimator of x_1 and the average of the scores of independent walks will be made to obtain the Monte Carlo approximation of x_1.

First the description of the original algorithm WE will be given as it is completely described in [12]. The WE Monte Carlo algorithm for computing the component u_{i_0} of the solution is given below:

Algorithm 1
Computing one component x_{i_0} of the solution $x_i, \, i = 1, \ldots n$

1. *Initialization* **Input** *initial data: the matrix B, the vector* **f***, the constant* γ *and the number of random trajectories N.*
2. *Preliminary calculations (preprocessing):*

   2.1. **Compute** *the matrix A using the parameter* $\gamma \in (0, 1]$:

   $$\{a_{ij}\}_{i,j=1}^n = \begin{cases} 1 - \gamma & \text{when} \quad i = j \\ -\gamma \frac{b_{ij}}{b_{ii}} & \text{when} \quad i \neq j \, . \end{cases}$$

   2.2. **Compute** *the vector asum:*

   $$asum(i) = \sum_{j=1}^n |a_{ij}| \quad for \quad i = 1, 2, \ldots, n.$$

   2.3. **Compute** *the transition probability matrix* $P = \{p_{ij}\}_{i,j=1}^n$, *where*

   $$p_{ij} = \frac{|a_{ij}|}{asum(i)}, \quad i = 1, 2, \ldots, n \quad j = 1, 2, \ldots, n \, .$$

3. **Set** $S := 0$.
4. **for** $k = 1$ **to** N **do** *(MC loop)*

   4.1 **set** $m := i_0$.
   4.2 **set** $S := S + b(m)$.

4.3. $test = 0$; $sign = 1$

4.4 **while** ($test \neq 0$ **do**:

    4.4.1. *generate an uniformly distributed random variable* $r \in (0, 1)$

    4.4.2. **if** $r \geq asum(m)$ **then** $test = 1$;

    4.4.3. **else** *chose the index* j *according to the probability density matrix* p_{ij};

    4.4.4. **set** $sign = sign * sign\{a_{mj}\}$; $m = j$;

    4.4.5 **update** $S := S + sign * b_m$;

    4.4.6. **endif**.

4.5 **endwhile**

5. **enddo**

After that the description the WE algorithm for computing all the components of the solution will be given, for more details see [12]. The case when all the matrix entrances are nonnegative will be considered.

Algorithm 2

Computing all components x_i, $i = 1, \ldots n$ **of the solution**

1. *Initialization* **Input** *initial data: the matrix* **B***, the vector* **f***, the constant* γ *and the number of random trajectories* N.

2. *Preliminary calculations (preprocessing):*

    2.1. **Compute** *the matrix* A *using the parameter* $\gamma \in (0, 1]$:

$$\{a_{ij}\}_{i,j=1}^n = \begin{cases} 1 - \gamma & \text{when } i = j \\ -\gamma \frac{b_{ij}}{b_{ii}} & \text{when } i \neq j. \end{cases}$$

    2.2. **Compute** *the vector asum:*

$$asum(i) = \sum_{j=1}^n |a_{ij}| \quad \text{for } i = 1, 2, \ldots, n.$$

    compute *the vector of initial probabilities,*

$$\pi_i = \frac{v_i}{\sum_{j=1}^m v_j}, \quad i = 1, 2, \ldots, m;$$

    2.3. **Compute** *the transition probability matrix* $P = \{p_{ij}\}_{i,j=1}^n$, *where*

$$p_{ij} = \frac{|a_{ij}|}{asum(i)}, \quad i = 1, 2, \ldots, n \quad j = 1, 2, \ldots, n.$$

3. **for** $i = 1$ **to** n **do**

3.1. $S(i) := 0$; $V(i) := 0$

3.2. **for** $k = 1$ **to** N **do** *(MC loop)*

3.1 **set** $m := rand(1 : n)$.

3.2 **set** $test := 0$; $m_1 := 0$;

3.3 **while** $(test \neq 1$ **do***:*

   3.3.1. $V(m) := V(m) + 1$; $m = m_1 + 1$; $l(m_1) = m$;

   3.3.2. **for** $q = 1, m_1$ **do***:*

      $S(l(q)) := S(l(q)) + b(l(q))$;

   3.3.3. **if** $r > asum(m)$, **then** $test = 1$;

      else *chose* j *according to the probability density matrix* $p_{i,j}$;

   3.3.4. **set** $m = j$.

   3.3.5. **endif**

3.4. **endwhile**

3.5. **enddo**

3.6. **for** $j = 1, n$ **do***:*

   $V(j) := \max\{1, V(j)\}$;

   $S(j) = \frac{S(j)}{V(j)}$.

From the Theorem 3 leads that the original algorithm can be optimized if we manage to balance the iteration matrix, which is completely described in the description of the new algorithm for computing all components of the solution of a linear system. Another optimization technique is to apply the sequential Monte Carlo (SMC) method for linear systems introduced by Halton [27–30]. Assume that one needs to solve the problem:

$$x = Ax + b, \quad A \in \mathbb{R}^{n \times n}, \quad b = (b_1, \dots, b_n)^t \in \mathbb{R}^{n \times 1}.$$

The main idea of the SMC method is quite simple but very efficient. It relies on an iterative use of the control variate method [12]. In the case of linear systems we need to compute the vector x solution to $x = Ax + b$. Obviously, if b is small, the variance of our algorithm is small and if $b = 0$, the variance is equal to zero. Because our problem is linear, we can easily construct a new system, which solution is the residual of our initial system, having a small b. Let $x^{(1)}$ the approximate solution of the linear system using algorithm 4.2. Then the residual $y = x - x^{(1)}$ is solution to the new equation $y = Ay + b - Ax^{(1)}$ where $b - Ax^{(1)}$ should be close to zero. We compute the approximation $y^{(1)}$ of this equation using our algorithm again and this leads to the approximation $x^{(2)} = x^{(1)} + y^{(1)}$ for the original equation. The same idea is used iteratively to obtain the estimation $x^{(k)}$ after k steps of the algorithm. One can prove (see [24], for instance) that if the number of samples N used to compute $x^{(k)}$ is large enough than its variance goes to zero geometrically fast with k. This type of convergence is usually denoted by zero-variance algorithms [6, 8, 17, 36–38, 40]. The algorithms of this kind can successfully compete with optimal deterministic algorithms based on Krilov subspace methods and the conjugate gradient methods [34]. But one should be careful, if the matrix-vector multiplication Ay is performed in a deterministic way, it needs $O(n^2)$ operations. It means that for a high quality WE algorithm combined with SMC one needs to have very few sequential steps [12].

Now we will introduce the new Monte Carlo algorithm for the computation a linear functional form $W(x)$ of the solution of a linear system with real coefficients. The original algorithm [12] algorithm is improved by choosing the appropriate values for the relaxation parameters which leads to the balancing of the iteration matrix which reduces the probable error according to the theorem for perfectly balanced matrix in the previous section. The idea is to compute scores for all the states (seen as new starting states) that are visited during a given trajectory. The initial equation is picked uniformly at random among the n equations. After that for each state i we define the total score $S(i)$ and the total number of visits $V(i)$ that are modified as soon as state i is visited during a walk. For a given trajectory, we store the visited states in a list l to compute the contributions to the score of these visited states. Finally, the solution x_i is approximated by the total score $S(i)$ divided by the total number of visits $V(i)$. If a state is never visited, we state $V(i) = 1$ anyway. If a state is visited more than once during the same walk, we usually only keep the contribution of the first visit to reduce the variance [42].

The matrices B and the right-hand side f are normalized to accelerate the convergence rate of the stochastic process. Special values of the relaxation parameter has been chosen, namely

$$\gamma_i = b_{ii}, \quad i = 1, \ldots, n. \tag{22}$$

Numerical experiments show that it leads to balancing of the iteration matrix A. For simplicity, we will consider the case when all the matrix entrances are nonnegative. We assume that $a_{ij} \geq 0$, for $i, j = 1, \ldots n$. The description of the new IWE Monte Carlo algorithm for computing all the components of the solution is given below:

Algorithm 3
Computing all components x_i, $i = 1, \ldots n$ of the solution

1. *Initialization* **Input** *initial data: the matrix B, the vector f, the constants*
   $\gamma_i = b_{ii}$, $i = 1, \ldots, n$ *and the number of random trajectories N.*
2. *Preliminary calculations (preprocessing):*

   2.1. **Compute** *the matrix A using the parameter $\gamma \in (0, 1]$:*

   $$\{a_{ij}\}_{i,j=1}^{n} = \begin{cases} 1 - b_{ii} & \text{when } i = j \\ -b_{ij} & \text{when } i \neq j. \end{cases}$$

3. **for** $i = 1$ **to** n **do**
3.1. $S(i) := 0; V(i) := 0$
3.2. **for** $k = 1$ **to** N **do**

   3.2.1 **set** $m := rand(1 : n)$.
   3.2.2 **set** $test := 0; m_1 := 0;$
   3.2.3 $V(m) := V(m) + 1; m = m_1 + 1; l(m_1) = m;$
   3.24 **for** $q = 1, m_1$ **do**: $S(l(q)) := S(l(q)) + b_{l(q)};$

3.5. **enddo**
3.6. **for** $j = 1, n$ **do**:
       $V(j) := \max\{1, V(j)\}$;
       $x_j = \frac{S(j)}{V(j)}$.
       enddo

4 Numerical Examples and Results

In order to check the accuracy of a computed solution $\hat{x}$, we compute the residual
[46, 50]

$$r := B\hat{x} - f \tag{23}$$

and "weighted residual" [39, 41]:

$$\rho := \frac{||r||}{||B|| \, ||\hat{x}||}. \tag{24}$$

The number of SMC iteration is N and the computational time t is measured in
seconds. In the tables below we present the values of the weighted residual. We
perform a comparison with the refined iterative Monte Carlo (MC) [8] and the original
"Walk on Equations" (WE) method which is completely described in [12]. For our
method we use the notation optimized "Walk on Equations" algorithm (IWE).

Example 1 We try to find the solutions x_1 and x_2 defined by the linear systems of
algebraic equations $Ax = b$, where the matrix B and the vectors b_1 and b_2 are:

$$B = \begin{pmatrix} 2 & -1 & 0 & 0 & 0 \\ -1 & 2 & -1 & 0 & 0 \\ 0 & -1 & 2 & -1 & 0 \\ 0 & 0 & -1 & 2 & -1 \\ 0 & 0 & 0 & -1 & 2 \end{pmatrix}, \quad f_1 = \begin{pmatrix} -1 \\ 1 \\ 0 \\ 0 \\ 1 \end{pmatrix}, \quad f_2 = \begin{pmatrix} -2 \\ 1 \\ 0 \\ 0 \\ 6 \end{pmatrix}. \tag{25}$$

The solutions are

$$x_1 = \begin{pmatrix} 0 \\ 1 \\ 1 \\ 1 \\ 1 \end{pmatrix}, \quad x_2 = \begin{pmatrix} 0 \\ 2 \\ 3 \\ 4 \\ 5 \end{pmatrix}. \tag{26}$$

Example 2 We try to find the solutions x_1 and x_2 defined by the linear systems of
algebraic equations $Ax = b$, where the matrix B and the vectors b_1 and b_2 are:

$$A = \begin{pmatrix} 5 & -1 & -1 & 0 & 0 & -1 & -1 \\ -1 & 5 & -1 & -1 & 0 & 0 & -1 \\ -1 & -1 & 5 & -1 & -1 & 0 & 0 \\ 0 & -1 & -1 & 5 & -1 & -1 & 0 \\ 0 & 0 & -1 & -1 & 5 & -1 & -1 \\ -1 & 0 & 0 & -1 & -1 & 5 & -1 \\ -1 & -1 & 0 & 0 & -1 & -1 & 5 \end{pmatrix}, \quad b_1 = \begin{pmatrix} 1 \\ 1 \\ 1 \\ 1 \\ 1 \\ 1 \\ 1 \end{pmatrix}, \quad b_2 = \begin{pmatrix} 4 \\ -2 \\ -1 \\ 0 \\ -1 \\ -2 \\ 4 \end{pmatrix}. \quad (27)$$

The solutions are

$$x_1 = \begin{pmatrix} 1 \\ 1 \\ 1 \\ 1 \\ 1 \\ 1 \\ 1 \end{pmatrix}, \quad x_2 = \begin{pmatrix} 1 \\ 0 \\ 0 \\ 0 \\ 0 \\ 0 \\ 1 \end{pmatrix}. \quad (28)$$

Example 3 Let B is a dense matrix 100×100 with elements in $[0, 1]$, and $b \in \mathbb{R}^{100}, b_i = 1, i = 1, \ldots, 100$.

Example 4 Let B is the matrix NOS4 from the Harwell-Boeing Collection, $f \in \mathbb{R}^{100}, f_i = 1, i = 1, \ldots, 100$. This particular matrix is taken from an application connected to finite element approximation of a problem describing a beam structure in constructive mechanics. It is real symmetric positive definite with 594 entries. Frobenius norm is 4.2, condition number (est.) is 2700 2-norm (est.) is 0.85, diagonal dominance no. This matrix is connected with problems in air transport and ecology. The structure of NOS4 is taken from [44] and it is shown on Fig. 1.

Example 5 Let B is positive definite sparse matrix 1000×1000 with random numbers in $(0, 1)$ and $f \in \mathbb{R}^{1000}, f_i = 1, i = 1, \ldots, 1000$.

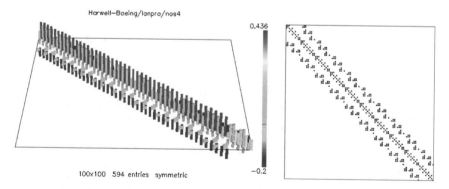

Harwell−Boeing/lanpro/nos4

100x100 594 entries symmetric

0.436

−0.2

Fig. 1 NOS4

Example 6 Let B is a dense matrix 5000×5000 with elements in $[0, 1]$, and $b \in \mathbb{R}^{5000}$, $b_i = 1$, $i = 1, \dots, 5000$.

Comparison of relative errors and computational times for a fixed number of iterations between RIMC, WE and IWE is given on tables below. The matrices B and the right-hand side b are normalized to accelerate the convergence rate of the stochastic process. Special values of the relaxation parameter γ has been chosen. Numerical experiments show that it leads to balancing of the iteration matrix A.

The numerical experiments show that for 5 dimensional case the WE and IWE produce very similar results—see Fig. 2, with slightly edge to the improved method with increasing the number of sequential Monte Carlo steps—see Table 1. It can be seen that for the 7 dimensional case the difference in the accuracy between the WE and OWE for a given number of iterations is 2–3 order for $N > 15$—see Table 2. It is worth mentioning that refined iterative Monte Carlo convergence is very slow except for the trivial solution x_1 of Example 2—see Fig. 3. For the 100 dimensional case the improved method produces much better results than WE and it is nearly 5–6 times faster—see Table 3.

The prior behavior of the proposed MC algorithm does not depend on the matrix density. The matrix NOS4 has only 5.9 average nonzeros per row and per column. The advantages of the algorithm hold for dense matrices. Also the difference in the

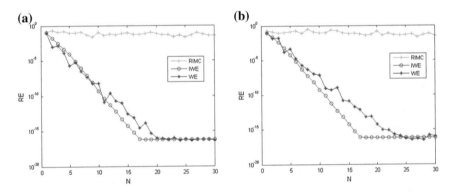

Fig. 2 Weighted residual for the matrix 5×5: **a** for x_1; **b** for x_2

Table 1 Weighted residual for the matrix $B \in \mathbb{R}^{5 \times 5}$

N	x_1						x_2					
	RIMC	t, s	WE	t, s	IWE	t, s	RIMC	t, s	WE	t, s	IWE	t, s
2	2.22e−01	0.01	1.01e−03	0.01	1.66e−02	0.01	1.42e−01	0.01	1.63e−02	0.01	2.35e−02	0.01
5	1.01e−01	0.02	2.03e−06	0.13	8.60e−05	0.02	1.30e−01	0.01	4.28e−04	0.12	6.07e−05	0.01
10	1.02e−01	0.03	6.18e−09	0.21	5.01e−10	0.04	8.81e−02	0.03	6.67e−08	0.20	5.17e−10	0.04
15	9.09e−02	0.05	2.47e−13	0.37	4.70e−15	0.08	6.31e−02	0.05	1.37e−11	0.35	5.62e−15	0.07
20	6.81e−02	0.09	9.62e−17	0.53	5.92e−17	0.11	4.11e−02	0.08	5.13e−15	0.51	6.12e−17	0.10
30	3.68e−02	0.14	4.44e−17	0.89	4.52e−17	0.19	3.55e−02	0.13	9.03e−17	0.86	7.13e−17	0.18

Table 2 Relative error for the matrix $B \in \mathbb{R}^{7 \times 7}$

N	x_1						x_2					
	RIMC	t, s	WE	t, s	IWE	t, s	RIMC	t, s	WE	t, s	IWE	t, s
2	2.28e−15	0.02	4.15e−02	0.11	1.00e−01	0.01	1.54e−01	0.003	4.63e−01	0.11	8.21e−02	0.01
5	7.89e−16	0.07	1.39e−02	0.23	2.29e−03	0.02	1.01e−01	0.01	9.93e−03	0.23	2.48e−03	0.02
10	8.07e−16	0.21	3.18e−06	0.68	1.24e−06	0.04	3.55e−02	0.04	1.43e−06	0.68	1.42e−06	0.05
15	7.45e−16	0.35	3.94e−08	1.11	2.55e−10	0.1	4.04e−02	0.08	6.17e−09	1.11	2.66e−10	0.09
20	6.66e−16	0.69	1.71e−10	2.53	7.78e−14	0.23	5.00e−02	0.14	1.40e−09	2.53	6.56e−14	0.16
30	5.79e−16	1.14	8.12e−15	3.69	8.32e−17	0.49	4.72e−02	0.24	1.53e−14	3.69	5.03e−17	0.29

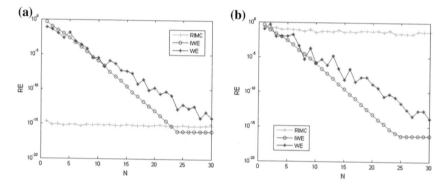

Fig. 3 Weighted residual for the matrix 7×7: **a** for x_1; **b** for x_2

Table 3 Weighted residual for the matrix $B \in \mathbb{R}^{100 \times 100}$ and $NOS4$

N	100×100						NOS4					
	RIMC	t, s	WE	t, s	IWE	t, s	RIMC	t, s	WE	t, s	IWE	t, s
2	5.877e−03	0.04	4.295e−02	0.47	2.388e−02	0.06	7.25e−02	0.05	4.18e−01	0.84	3.02e−03	0.08
5	3.729e−03	0.21	1.662e−01	1.23	2.414e−04	0.28	5.45e−02	0.22	4.14e−01	2.37	3.07e−05	0.24
10	2.616e−03	0.54	2.491e−05	2.68	1.565e−08	0.59	4.32e−02	0.56	5.94e−03	5.31	7.46e−08	0.61
15	2.726e−03	0.88	3.064e−07	4.11	3.201e−10	0.89	3.52e−02	0.78	2.41e−06	9.1	1.21e−10	0.89
20	2.134e−03	1.24	8.342e−08	8.53	2.611e−13	1.21	3.19e−02	1.11	3.33e−09	13.5	1.02e−13	1.13
30	1.722e−03	2.54	9.511e−11	19.69	2.682e−16	2.29	1.83e−02	2.15	3.66e−12	24.6	1.11e−16	1.92

accuracy for a fixed number of iterations is 3–5 order—see Fig. 4. Therefore IWE has strong advantage in terms of superior accuracy with increasing the dimensionality of the system. Also the advantage of the new algorithm is clearly seen in terms of smaller CPU time. We compare our results with the optimal deterministic preconditioned conjugate gradient (PCG) method [26, 35, 46]. We want to solve the linear system of equations $Bx = f$ by means of the PCG iterative method. The input arguments are the matrix B, which in our case is the square large matrix $NOS4$ taken from the well-known Harwell-Boeing Collection. B should be symmetric and positive definite. In our implementation we use the following parameters: f is the right-hand side vector; *tol* is the required relative tolerance for the residual error, namely

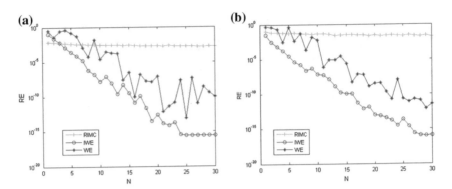

Fig. 4 Weighted residual for the matrix: **a** 100×100; **b** NOS4

$f - Bx$. The iteration stops if $\| f - Bx \| \leq tol * \| f \|$; $maxit$ is the maximum allowable number of iterations; $M = M_1 * M_2$ is the (left) preconditioning matrix, so that the iteration is (theoretically) equivalent to solving by PCG $Px = M f$, with $P = M B$; x_0 is the initial guess. The results presented on Fig. 5 show that the convergence for the WE Monte Carlo is better: the curve presenting the residual with the number of iterations is smoother and goes down to 10^{-6} or 10^{-7}, while the curve presenting the PCG achieves an accuracy of 10^{-3}. Let us stress on the fact that for the experiments presented the needed accuracy is set to $tol = 10^{-8}$. This result can be explained in the following way. The success of the PCG method is based on the fact that it operates in optimal Krilov subspaces. Our WE algorithm also operates in optimal Krilov subspaces. The difference is that in PCG one actually solves an optimization problem, and if there are local minima close to the global one, the process may go to the local minimum, which was the case in the example shown on Fig. 5. Our approach is free of such an optimization procedure, that is why the results are better. The particular matrix (NOS4) seems to be a hard test for the PCG. We do not generalize this fact, because the success depends on the particular functional that should be minimized in the PCG method. This is a matter of further analysis which is not trivial. One can not guarantee that such an effect happens for every matrix, but there are cases in which the IWE and WE Monte Carlo methods perform better than the widely used PCG.

For larger dimensions the advantage of IWE over WE is even more pronounced. For the 1000 dimensional sparse matrix IWE produces the same relative error after 15 sequential Monte Carlo steps, as the original algorithm for 30 sequential iterations—see Table 4. For 5000 dimensional dense matrix after 30 iterations WE has accuracy of 10^{-9} while IWE gives accuracy of 10^{-16}—see Fig. 6. Also the time for the IWE is 15 times better than WE. The choice of relaxation parameters leads to the balancing of iteration matrix which reduces the probable error according to the theorem for the perfectly balanced matrix.

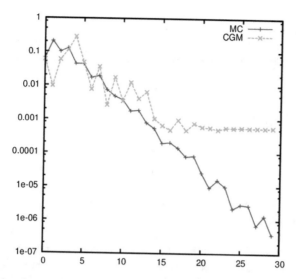

Fig. 5 Comparison between PCG and WE for $NOS4$ from Harwell-Boeing

Table 4 Relative error for the matrix $B \in \mathbb{R}^{1000 \times 1000}$ and $B \in \mathbb{R}^{5000 \times 5000}$

N	1000×1000						5000×5000					
	RIMC	t, s	WE	t, s	IWE	t, s	RIMC	t, s	WE	t, s	IWE	t, s
2	1.212e−03	2.17	1.035e−01	0.75	1.064e−02	0.11	5.43e−03	10.05	4.30e−02	3.95	2.93e−02	0.15
5	7.397e−04	13.1	4.621e−03	2.34	3.468e−05	0.45	3.87e−03	60.2	1.21e−01	13.3	1.82e−04	0.9
10	5.343e−04	33.3	5.819e−05	6.3	2.545e−10	0.97	2.86e−03	130.5	2.30e−05	32.3	1.23e−07	2.4
15	4.420e−04	65	9.309e−07	13.5	3.823e−13	2.3	2.36e−03	310.7	6.48e−09	67.8	1.83e−10	5.1
20	3.554e−04	169	8.670e−10	36	7.731e−16	6.7	1.94e−03	811	3.20e−09	171.5	1.05e−14	11.1
30	3.063e−04	357	4.436e−14	87	7.325e−16	17	1.70e−03	2135	1.12e−07	418.6	2.48e−16	25.2

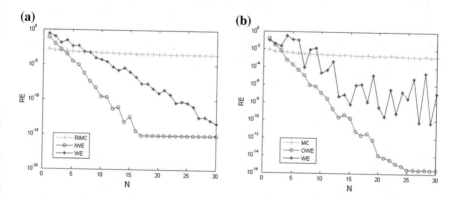

Fig. 6 Weighted residual for the matrix: **a** 1000×1000; **b** 5000×5000

The advantages of the proposed MC algorithm can be observed especially for larger matrix size. Experiments show that for larger dimensions the improvements leads to lower relative errors for small number of SMC iterations for IWE.

5   Concluding Remarks

A new Monte Carlo algorithm based on "Walk on Equations" Monte Carlo algorithm for solving linear algebra problems is presented and studied. The algorithm can be used for evaluating all the components of the solution of a linear system. A number of numerical experiments are performed. The analysis of the results show that the proposed IWE Monte Carlo algorithm combined with special choice of the relaxation parameter and the sequential Monte Carlo for computing all the components of the solution converges much faster than the original method and refined iterative Monte Carlo method; this is true for matrices of different size and the effect is much bigger for larger matrices. For some matrices like NOS4, the improved "Walk on Equations" Monte Carlo method gives better results than the preconditioned conjugate gradient method. Due to the optimization techniques the new method gives superior results to the standard "Walk on Equations" method and it is established as one of the fastest and accurate Monte Carlo algorithm for solving systems of linear algebraic equations.

Acknowledgements The authors are supported by the Bulgarian National Science Fund under Projects DN 12/5-2017 "Efficient Stochastic Methods and Algorithms for Large-Scale Problems", DN 12/4-2017 "Advanced Analytical and Numerical Methods for Nonlinear Differential Equations with Applications in Finance and Environmental Pollution" and Bilateral Project Bulgaria-Russia DNTS 02/12-2018 "Development and investigation of finite-difference schemes of higher order of accuracy for solving applied problems of fluid and gas mechanics, and ecology", by the National Scientific Program "Information and Communication Technologies for a Single Digital Market in Science, Education and Security" (ICTinSES), contract No DO1–205/23.11.2018, financed by the Ministry of Education and Science and by the Grant BG05M2OP001-1.001-0003, financed by the Science and Education for Smart Growth Operational Program (2014–2020) and co-financed by the EU through the European structural and Investment funds.

References

1. Alexandrov, V., Atanassov, E., Dimov, I.: Parallel quasi-Monte Carlo methods for linear algebra problems. Monte Carlo Methods Appl. **10**(3–4), 213–219 (2004)
2. Asmussen, S., Glynn, P.W.: Stochastic Simulation. Springer, New York (2007)
3. Curtiss, J.H.: Monte Carlo methods for the iteration of linear operators. J. Math. Phys. **32**(4), 209–232 (1954)
4. Curtiss, J.H.: A theoretical comparison of the efficiencies of two classical methods and a Monte Carlo method for computing one component of the solution of a set of linear algebraic equations. In: Proceedings of the Symposium on Monte Carlo Methods, pp. 191–233. Wiley (1956)

5. Densmore, J.D., Larsen, E.W.: Variational variance reduction for particle transport eigenvalue calculations using Monte Carlo adjoint simulation. J. Comput. Phys. **192**(2), 387–405 (2003)
6. Dimov, I.T.: Minimization of the probable error for some Monte Carlo methods. In: Proceedings of the International Conference on Mathematical Modeling and Scientific Computation, Albena, Bulgaria, Sofia, pp. 159–170. Publishing House of the Bulgarian Academy of Sciences (1991)
7. Dimov, I.: Optimal Monte Carlo algorithms. In: Proceedings IEEE John Vincent Atanasoff 2006 International Symposium on Modern Computing, Oct 2006, Sofia, Bulgaria. IEEE, Los Alamitos, California, pp. 125–131 (2006)
8. Dimov, I.: Monte Carlo Methods for Applied Scientists, 291 pp. World Scientific, New Jersey, London, Singapore (2008)
9. Dimov, I.T., Alexandrov, V.: A new highly convergent Monte Carlo method for matrix computations. Math. Comput. Simul. **47**, 165–181 (1998)
10. Dimov, I., Alexandrov, V., Karaivanova, A.: Parallel resolvent Monte Carlo algorithms for linear algebra problems. J. Math. Comput. Simul. **55**, 25–35 (2001)
11. Dimov, I., Gurov, T.: Monte Carlo algorithm for solving integral equations with polynomial non-linearity. Parallel Implement. Pliska (Studia Mathematica Bulgarica) **13**, 117–132 (2000)
12. Dimov, I.T., Maire, S., Sellier, J.M.: A new walk on equations Monte Carlo method for linear algebraic problems. Appl. Math. Model. **39**(15), 4494–4510 (2015)
13. Dimov, I.T., Dimov, T., Gurov, T.: A new iterative Monte Carlo approach for inverse matrix problem. J. Comput. Appl. Math. **92**, 15–35 (1998)
14. Dimov, I.T., Karaivanova, A.N.: Iterative Monte Carlo algorithms for linear algebra problems. In: First Workshop on Numerical Analysis and Applications, Rousse, Bulgaria, 24–27 June 1996. Numerical Analysis and Its Applications. Springer Lecture Notes in Computer Science, ser. 1196, pp. 150–160
15. Dimov, I.T., Karaivanova, A.: Parallel computations of eigenvalues based on a Monte Carlo approach. J. Monte Carlo Method Appl. **4**(1), 33–52 (1998)
16. Dimov, I.T., Karaivanova, A.: A power method with Monte Carlo iterations. In: Iliev, O., Kaschiev, M., Sendov, Bl., Vassilevski, P. (eds.) Recent Advances in Numerical Methods and Applications, pp. 239–247. World Scientific, Singapore (1999)
17. Dimov, I.T., Philippe, B., Karaivanova, A., Weihrauch, C.: Robustness and applicability of Markov chain Monte Carlo algorithms for eigenvalue problems. Appl. Math. Model. **32**, 1511–1529 (2008)
18. Dimov, I.T., Tonev, O.: Performance analysis of Monte Carlo algorithms for some models of computer architectures. In: Sendov, Bl., Dimov, I.T. (eds.) International Youth Workshop on Monte Carlo Methods and Parallel Algorithms—Primorsko, pp. 91–95. World Scientific, Singapore (1990)
19. Dimov, I.T., Tonev, O.: Monte Carlo algorithms: performance analysis for some computer architectures. J. Comput. Appl. Math. **48**, 253–277 (1993)
20. Ermakov, S.M., Mikhailov, G.A.: Statistical Modeling. Nauka, Moscow (1982)
21. Fernndez, C., et al.: Controller design for tracking paths in nonlinear biochemical processes. In: 2016 IEEE Biennial Congress of Argentina (ARGENCON). IEEE (2016)
22. Fernndez, M.C., et al.: A new approach for nonlinear multivariable fed-batch bioprocess trajectory tracking control. Autom. Control Comput. Sci. **52**(1), 13–24 (2018)
23. Forsythe, G.E., Leibler, R.A.: Matrix inversion by a Monte Carlo method. MTAC **4**, 127–129 (1950)
24. Gobet, E., Maire, S.: Sequential control variates for functionals of Markov processes. SIAM J. Numer. Anal. **43**, 1256–1275 (2005)
25. Golub, G.H., Van Loon, C.F.: Matrix Computations, 3rd edn. Johns Hopkins University Press, Baltimore (1996)
26. Golub, G.H., Ye, Q.: Inexact preconditioned conjugate gradient method with inner-outer iteration. SIAM J. Sci. Comput. **21**(4), 1305 (1999). https://doi.org/10.1137/S1064827597323415
27. Halton, J.: Sequential Monte Carlo. In: Proceedings of the Cambridge Philosophical Society, vol. 58, pp. 57–78 (1962)

28. Halton, J.: Sequential Monte Carlo, University of Wisconsin, Madison, Mathematics Research Center Technical Summary Report No. 816, 38 pp. (1967)
29. Halton, J., Zeidman, E.A.: Monte Carlo integration with sequential stratification, University of Wisconsin, Madison, Computer Sciences Department Technical Report No. 61, 31 pp. (1969)
30. Halton, J.: Sequential Monte Carlo for linear systems—a practical summary. Monte Carlo Methods Appl. **14**, 1–27 (2008)
31. Hammersley, J.M., Handscomb, D.C.: Monte Carlo Methods. Wiley, New York, London, Sydney, Methuen (1964)
32. Jeng, C., Tan, K.: Solving systems of linear equations with relaxed Monte Carlo method. J. Supercomput. **22**, 113–123 (2002)
33. Ji, H., Mascagni, M., Li, Y.: Convergence analysis of Markov chain Monte Carlo linear solvers using Ulam-von Neumann algorithm. SIAM J. Numer. Anal. **51**(4), 2107–2122 (2013)
34. Kantorovich, L.W., Krylov, V.I.: Approximate Methods of Higher Analysis. Interscience, New York (1964)
35. Knyazev, A.V., Lashuk, I.: Steepest descent and conjugate gradient methods with variable preconditioning. SIAM J. Matrix Anal. Appl. **29**(4), 1267 (2008). https://doi.org/10.1137/060675290
36. Kollman, C., Baggerly, K., Cox, D., Picard, R.: Adaptive importance sampling on discrete Markov chains. Ann. Appl. Probab. **9**(2), 391–412 (1999)
37. L'Ecuyer, P., Blanchet, J.H., Tuffin, B., Glynn, P.W.: Asymptotic robustness of estimators in rare-event simulation. ACM Trans. Model. Comput. Simul. **20**(1), Article 6 (2010)
38. L'Ecuyer, P., Tuffin, B.: Approximate zero-variance simulation. In: Proceedings of the 2008 Winter Simulation Conference, pp. 170–181. IEEE Press (2008)
39. Linear systems. www.math.umd.edu/~petersd/466/linsysterrn.pdf
40. Maire, S.: Reducing variance using iterated control variates. J. Stat. Comput. Simul. **73**(1), 1–29 (2003)
41. MATH 3795 Lecture 6. Sensitivity of the Solution of a Linear System. http://www.math.uconn.edu/~leykekhman/courses/MATH3795/Lectures/Lecture/6
42. Medvedev, I., Mikhailov, G.: A new criterion for finiteness of weight estimator variance in statistical simulation. In: Keller, A., Heinrich, S., Niederreiter, H. (eds.) Monte Carlo and Quasi-Monte Carlo Methods, 2006, pp. 561–576 (2008)
43. Mikhailov, G.A.: Optimization of Weight Monte Carlo methods. Nauka, Moscow (1987)
44. NOS4: Lanczos with partial reorthogonalization. Finite element approximation to a beam structure. http://math.nist.gov/MatrixMarket/data/Harwell-Boeing/lanpro/nos4.html
45. Rosca, N.: A new Monte Carlo estimator for systems of linear equations. Studia Univ. Babes–Bolyai Math. **LI**(2), pp. 97–107 (2006)
46. Saad, Y.: Iterative Methods for Sparse Linear Systems, 2nd edn., p. 195. Society for Industrial and Applied Mathematics, Philadelphia, PA (2003). ISBN 978-0-89871-534-7
47. Shyamkumar, N., Banerjee, S., Lofgren, P.: Sublinear estimation of a single element in sparse linear systems. In: 2016 54th Annual Allerton Conference on Communication, Control, and Computing (Allerton). IEEE (2016)
48. Sobol, I.M.: Monte Carlo Numerical Methods. Nauka, Moscow (1973)
49. Straburg, J., Alexandrov, V.N.: A Monte Carlo approach to sparse approximate inverse matrix computations. Procedia Comput. Sci. **18**, 2307–2316 (2013)
50. Westlake, J.R.: A Handbook of Numerical Matrix Inversion and Solution of Linear Equations. Wiley, New York, London, Sydney (1968)

Author Index

A
Angelova, Jordanka A., 105
Apostolov, Stoyan, 215
Atanassov, Atanas V., 105
Atanassova, Vassia, 155, 181

B
Bourne, Paul, 1

D
Dörpinghaus, Jens, 45
Dimitrov, Yuri, 215
Dimov, Ivan, 215
Dobrinkova, Nina, 17, 31

F
Fidanova, Stefka, 61, 83, 133
Filipov, Stefan M., 105
Filippova, Tatiana F., 121

G
Georgieva, Rayna, 215
Gospodinov, Ivan D., 105

I
Ikonomov, Nikolay, 215

K
Kirilov, Leoneed, 133

L
Luque, Gabriel, 61

M
Mikhov, Rossen, 133
Mucherino, Antonio, 147
Myasnichenko, Vladimir, 133

P
Palmer, Neville, 1
Paprzycki, Marcin, 61

R
Roeva, Olympia, 61, 83

S
Schrader, Rainer, 45
Sdobnyakov, Nickolay, 133
Skrabala, Jan, 1
Stefanov, Stefan, 17
Stoykova, Velislava, 201

T
Todorov, Venelin, 215
Traneva, Velichka, 155, 181
Tranev, Stoyan, 155, 181

© Springer Nature Switzerland AG 2020
S. Fidanova (ed.), *Recent Advances in Computational Optimization*,
Studies in Computational Intelligence 838,
https://doi.org/10.1007/978-3-030-22723-4

Printed in the United States
By Bookmasters